The comical history of the states and empires of the worlds of the moon and sun written in French by Cyrano Bergerac; and newly Englished by A. Lovell ... (1687)

Archibald Lovell

The comical history of the states and empires of the worlds of the moon and sun written in French by Cyrano Bergerac ; and newly Englished by A. Lovell ...
Histoire comique des états et empires du soleil.
Cyrano de Bergerac, 1619-1655.
Lovell, Archibald.
Translation of: Histoire comique des états et empires du soleil.
Each volume has special t.p.
Pages 45-50 lacking in the filmed copy. Pages 40-65 photographed from Yale University Library copy and inserted at the end.
2 v. :
London : Printed for Henry Rhodes ..., 1687.
Arber's Term cat. / II 166
Wing / C7717
English
Reproduction of the original in the Harvard University Library

Early English Books Online (EEBO) Editions

Imagine holding history in your hands.

Now you can. Digitally preserved and previously accessible only through libraries as **Early English Books Online**, this rare material is now available in single print editions. Thousands of books written between 1475 and 1700 and ranging from religion to astronomy, medicine to music, can be delivered to your doorstep in individual volumes of high-quality historical reproductions.

We have been compiling these historic treasures for more than 70 years. Long before such a thing as "digital" even existed, ProQuest founder Eugene Power began the noble task of preserving the British Museum's collection on microfilm. He then sought out other rare and endangered titles, providing unparalleled access to these works and collaborating with the world's top academic institutions to make them widely available for the first time. This project furthers that original vision.

These texts have now made the full journey -- from their original printing-press versions available only in rare-book rooms to online library access to new single volumes made possible by the partnership between artifact preservation and modern printing technology. A portion of the proceeds from every book sold supports the libraries and institutions that made this collection possible, and that still work to preserve these invaluable treasures passed down through time.

This is history, traveling through time since the dawn of printing to your own personal library.

Initial Proquest EEBO Print Editions collections include:

Early Literature

This comprehensive collection begins with the famous Elizabethan Era that saw such literary giants as Chaucer, Shakespeare and Marlowe, as well as the introduction of the sonnet. Traveling through Jacobean and Restoration literature, the highlight of this series is the Pollard and Redgrave 1475-1640 selection of the rarest works from the English Renaissance.

Early Documents of World History

This collection combines early English perspectives on world history with documentation of Parliament records, royal decrees and military documents that reveal the delicate balance of Church and State in early English government. For social historians, almanacs and calendars offer insight into daily life of common citizens. This exhaustively complete series presents a thorough picture of history through the English Civil War.

Historical Almanacs

Historically, almanacs served a variety of purposes from the more practical, such as planting and harvesting crops and plotting nautical routes, to predicting the future through the movements of the stars. This collection provides a wide range of consecutive years of "almanacks" and calendars that depict a vast array of everyday life as it was several hundred years ago.

Early History of Astronomy & Space

Humankind has studied the skies for centuries, seeking to find our place in the universe. Some of the most important discoveries in the field of astronomy were made in these texts recorded by ancient stargazers, but almost as impactful were the perspectives of those who considered their discoveries to be heresy. Any independent astronomer will find this an invaluable collection of titles arguing the truth of the cosmic system.

Early History of Industry & Science

Acting as a kind of historical Wall Street, this collection of industry manuals and records explores the thriving industries of construction; textile, especially wool and linen; salt; livestock; and many more.

Early English Wit, Poetry & Satire

The power of literary device was never more in its prime than during this period of history, where a wide array of political and religious satire mocked the status quo and poetry called humankind to transcend the rigors of daily life through love, God or principle. This series comments on historical patterns of the human condition that are still visible today.

Early English Drama & Theatre

This collection needs no introduction, combining the works of some of the greatest canonical writers of all time, including many plays composed for royalty such as Queen Elizabeth I and King Edward VI. In addition, this series includes history and criticism of drama, as well as examinations of technique.

Early History of Travel & Geography

Offering a fascinating view into the perception of the world during the sixteenth and seventeenth centuries, this collection includes accounts of Columbus's discovery of the Americas and encompasses most of the Age of Discovery, during which Europeans and their descendants intensively explored and mapped the world. This series is a wealth of information from some the most groundbreaking explorers.

Early Fables & Fairy Tales

This series includes many translations, some illustrated, of some of the most well-known mythologies of today, including Aesop's Fables and English fairy tales, as well as many Greek, Latin and even Oriental parables and criticism and interpretation on the subject.

Early Documents of Language & Linguistics

The evolution of English and foreign languages is documented in these original texts studying and recording early philology from the study of a variety of languages including Greek, Latin and Chinese, as well as multilingual volumes, to current slang and obscure words. Translations from Latin, Hebrew and Aramaic, grammar treatises and even dictionaries and guides to translation make this collection rich in cultures from around the world.

Early History of the Law

With extensive collections of land tenure and business law "forms" in Great Britain, this is a comprehensive resource for all kinds of early English legal precedents from feudal to constitutional law, Jewish and Jesuit law, laws about public finance to food supply and forestry, and even "immoral conditions." An abundance of law dictionaries, philosophy and history and criticism completes this series.

Early History of Kings, Queens and Royalty

This collection includes debates on the divine right of kings, royal statutes and proclamations, and political ballads and songs as related to a number of English kings and queens, with notable concentrations on foreign rulers King Louis IX and King Louis XIV of France, and King Philip II of Spain. Writings on ancient rulers and royal tradition focus on Scottish and Roman kings, Cleopatra and the Biblical kings Nebuchadnezzar and Solomon.

Early History of Love, Marriage & Sex

Human relationships intrigued and baffled thinkers and writers well before the postmodern age of psychology and self-help. Now readers can access the insights and intricacies of Anglo-Saxon interactions in sex and love, marriage and politics, and the truth that lies somewhere in between action and thought.

Early History of Medicine, Health & Disease

This series includes fascinating studies on the human brain from as early as the 16th century, as well as early studies on the physiological effects of tobacco use. Anatomy texts, medical treatises and wound treatment are also discussed, revealing the exponential development of medical theory and practice over more than two hundred years.

Early History of Logic, Science and Math

The "hard sciences" developed exponentially during the 16th and 17th centuries, both relying upon centuries of tradition and adding to the foundation of modern application, as is evidenced by this extensive collection. This is a rich collection of practical mathematics as applied to business, carpentry and geography as well as explorations of mathematical instruments and arithmetic; logic and logicians such as Aristotle and Socrates; and a number of scientific disciplines from natural history to physics.

Early History of Military, War and Weaponry

Any professional or amateur student of war will thrill at the untold riches in this collection of war theory and practice in the early Western World. The Age of Discovery and Enlightenment was also a time of great political and religious unrest, revealed in accounts of conflicts such as the Wars of the Roses.

Early History of Food

This collection combines the commercial aspects of food handling, preservation and supply to the more specific aspects of canning and preserving, meat carving, brewing beer and even candy-making with fruits and flowers, with a large resource of cookery and recipe books. Not to be forgotten is a "the great eater of Kent," a study in food habits.

Early History of Religion

From the beginning of recorded history we have looked to the heavens for inspiration and guidance. In these early religious documents, sermons, and pamphlets, we see the spiritual impact on the lives of both royalty and the commoner. We also get insights into a clergy that was growing ever more powerful as a political force. This is one of the world's largest collections of religious works of this type, revealing much about our interpretation of the modern church and spirituality.

Early Social Customs

Social customs, human interaction and leisure are the driving force of any culture. These unique and quirky works give us a glimpse of interesting aspects of day-to-day life as it existed in an earlier time. With books on games, sports, traditions, festivals, and hobbies it is one of the most fascinating collections in the series.

bibliolife
old books. new life.

The BiblioLife Network

This project was made possible in part by the BiblioLife Network (BLN), a project aimed at addressing some of the huge challenges facing book preservationists around the world. The BLN includes libraries, library networks, archives, subject matter experts, online communities and library service providers. We believe every book ever published should be available as a high-quality print reproduction; printed on-demand anywhere in the world. This insures the ongoing accessibility of the content and helps generate sustainable revenue for the libraries and organizations that work to preserve these important materials.

The following book is in the "public domain" and represents an authentic reproduction of the text as printed by the original publisher. While we have attempted to accurately maintain the integrity of the original work, there are sometimes problems with the original work or the micro-film from which the books were digitized. This can result in minor errors in reproduction. Possible imperfections include missing and blurred pages, poor pictures, markings and other reproduction issues beyond our control. Because this work is culturally important, we have made it available as part of our commitment to protecting, preserving, and promoting the world's literature.

GUIDE TO FOLD-OUTS MAPS and OVERSIZED IMAGES

The book you are reading was digitized from microfilm captured over the past thirty to forty years. Years after the creation of the original microfilm, the book was converted to digital files and made available in an online database.

In an online database, page images do not need to conform to the size restrictions found in a printed book. When converting these images back into a printed bound book, the page sizes are standardized in ways that maintain the detail of the original. For large images, such as fold-out maps, the original page image is split into two or more pages

Guidelines used to determine how to split the page image follows:

- Some images are split vertically; large images require vertical and horizontal splits.
- For horizontal splits, the content is split left to right.
- For vertical splits, the content is split from top to bottom.
- For both vertical and horizontal splits, the image is processed from top left to bottom right.

Printed for Henry Rhodes, next the Swan Tavern in Fleet street.

THE
Comical HISTORY
OF THE
STATES
AND
EMPIRES
OF THE
WORLDS
OF THE
𝕸𝖔𝖔𝖓 𝖆𝖓𝖉 𝕾𝖚𝖓.

Written in *French* by *Cyrano Bergerac.*

And newly Englished by *A. Lovell,* A.M.

LONDON,

Printed for *Henry Rhodes*, next door to the *Swan-Tavern*, near *Bride-Lane*, in *Fleet-Street*, 1687.

LICENSED,

May 30.
1686.

RO. L'ESTRANGE

THE TRANSLATOR TO THE READER.

IT is now *Seven and Twenty Years, since the Moon appeared first Historically on the* English *Horizon: And let it not seem strange, that she should have retained Light and Brightness so long here, without Re-vovation; when we find by Experience, that in the Heavens, she never fails once a Month to change and shift her Splendor. For it is the Excellency of Art, to represent Nature, even in her absence; and this being a Piece done to the Life, by one that had the advantage of the true Light, as well as the Skill of Drawing, in this kind, to Perfection; he left so good an Original, which was so well Copied by another Hand, that the Picture*

A 2 *might*

The Translator

might have served for many Years more, to have given the Lovers of the Moon, a sight of their Mistress, even in the darkest Nights; and when she was retired to put on a clean Smock in Phœbus *his Apartment; if they had been so curious, as to have encouraged the Exposers.*

However, Reader, you have now a second View of her, and that under the same Cover with the Sun too, which is very rare; since these two were never seen before in Conjunction. Yet I would have none be afraid, that their Eyes being dazled with the glorious Light of the Sun, they should not see her; for Fancy will supply the Weakness of the Organ, and Imagination, by the help of this Mirrour, will not fail to discover them both; though Cynthia *lye hid under* Apollo's *shining Mantle. And so much for the Luminaries.*

Now as to the Worlds, which, with Analogy to ours below, I may call the Old and New; that of the Moon having been discovered, tho imperfectly, by others, but the Sun owing its Discovery wholly to our Author: I make no doubt, but the Ingenious Reader will find in both, so extraordinary and surprizing Rarities, as well Natural, Moral, as Civil; that if he be not as yet sufficiently disgusted with this lower World,
(which

to the Reader.

(which I am sure some are) to think of making a Voyage thither, as our Author has done; he will at least be pleased with his Relations. Nevertheless, since this Age produces a great many bold Wits, that shoot even beyond the Moon, and cannot endure, (no more than our Author) to be stinted by Magisterial Authority, and to believe nothing but what Gray-headed Antiquity gives them leave: It's pity some soaring Virtuoso, instead of Travelling into France, does not take a flight up to the Sun; and by new Observations supply the defects of its History; occasioned not by the Negligence of our Witty French Author, but by the accursed Plagiary of some rude Hand, that in his Sickness, rifled his Trunks, and stole his Papers, as he himself complains.

Let some venturous Undertaker auspiciously attempt it then; and if neither of the two Universities, Gresham-Colledge, nor Greenwich-Observatory can furnish him with an Instrument of Conveyance; let him try his own Invention, or make use of our Author's Machine: For our Loss is, indeed, so great, that one would think, none but the declared Enemy of Mankind, would have had the Malice, to purloyn and stifle those rare Discoveries, which our Author made in the Province of the Solar Philosophers; and which un-

The Translator, &c.

doubtedly would have gone far, as to the settleing our Sublunary Philosophy, which, as well as Religion, is lamentably rent by Sects and Whimseys; and have convinced us, perhaps, that in our present Doubts and Perplexities, a little more, or a little less of either, would better serve our Turns, and more content our Minds.

THE

THE
Comical History
OF THE
STATES and *EMPIRES*
OF THE
WORLD
OF THE
MOON.

Written in *French* by *Cyrano Bergerac*.

And now *Englished* by *A. Lovell.* A. M.

LONDON,

Printed for *Henry Rhodes,* next door to the *Swan-Tavern,* near *Bride-Lane* in *Fleet-Street,* 1687.

THE Comical HISTORY OF THE STATE AND EMPIRE OF THE WORLD OF THE MOON.

I Had been with some Friends at *Clamard*, a House near *Paris*, and magnificently Entertain'd there by *Monsieur de Guigy*, the Lord of it; when upon

our

our return home, about Nine of the Clock at Night, the Air serene, and the Moon in the Full, the Contemplation of that bright Luminary, furnished us with such variety of Thoughts, as made the way seem shorter than, indeed, it was. Our Eyes being fixed upon that stately Planet, every one spoke what he thought of it: One would needs have it be a Garret Window of Heaven; another presently affirmed, That it was the Pan whereupon *Diana* smoothed *Apollo*'s Bands; whilst another was of Opinion, That it might very well be the Sun himself, who putting his Locks up under his Cap at Night, peeped through a hole, to observe what was doing in the World, during his absence: And for my part, Gentlemen, said I, that I may put in for a share, and guess with the rest; not to amuse my self with those curious Notions, wherewith you tickle and spur on slow-paced Time; I believe, that the Moon is a World like ours, to which this of ours serves likewise for a Moon. This was received with the general Laughter of the Company. And perhaps, said I, (Gentlemen) just so they laugh now in the Moon, at some who maintain, That this Globe, where we are, is a World. But I'd as good have said nothing, as have alledged to them, That a great many Learned

Men

World of the Moon.

Men had been of the same Opinion; for that only made them laugh the faster. However, this thought, which because of its boldness suted my Humor, being confirmed by Contradiction, sunk so deep into my mind, that during the rest of the way, I was big with Definitions of the Moon, which I could not be delivered of: Insomuch that by striving to verifie this Comical Fancy, by Reasons of appearing weight, I had almost perswaded my self already of the truth on't; when a Miracle, Accident, Providence, Fortune, or what, perhaps, some may call Vision, others Fiction, Whimsey, or (if you will) Folly, furnished me with an occasion, that engaged me into this Discourse. Being come home, I went up into my Closet, where I found a Book open upon the Table, which I had not put there. It was a piece of *Cardanus*; and though I had no design to read in it, yet I fell at first sight, as by force, exactly upon a Passage of that Philosopher, where he tells us, That Studying one evening by Candle-light, he perceived Two tall old Men, enter in through the door that was shut, who after many questions that he put to them, made him answer, That they were Inhabitants of the Moon, and thereupon immediately disappeared. I was so surprised, not only to see a Book get thither of it self; but also because of the nicking of the Time so patly, and of the Page, at which it lay open, that I looked upon that Concatenation of Accidents, as a Revelation, discovering to Mortals, that the Moon is a World. How! said I to my self, having just now talked of a thing, can a Book, which, perhaps, is the only Book in the World, that treats of that matter so particularly, fly down from the Shelf upon my Table, become capable of Reason, in opening so exactly at the place of so strange an adventure; force my Eyes in a manner to look upon it, and then to suggest to my fancy the Reflexions, and to my Will the Designs which I hatch: Without doubt, continued I, the Two old Men, who appeared to that famous Philosopher, are the very same who have taken down my Book, and opened it at that Page, to save themselves the labour of making to me the Harangue, which they made to *Cardan*. But, added I, I cannot be resolved of this Doubt, unless I mount up thither: And why not? said I instantly to my self. *Promethus* heretofore went up to Heaven, and stole fire from thence. Have not I as much Boldness as he? And why should not I, then, expect as favourable a Success?

After

After these sudden starts of Imagination, which may be termed, perhaps, the Ravings of a violent Feaver, I began to conceive some hopes of succeeding in so fair a Voyage: Insomuch that to take my measures aright, I shut my self up in a solitary Country-house; where having flattered my fancy with some means, proportionated to my design, at length I set out for Heaven in this manner.

I planted my self in the middle of a great many Glasses full of Dew, tied fast about me; upon which the Sun so violently darted his Rays, that the Heat which attracted them, as it does the thickest Clouds, carried me up so high, that at length I found my self above the middle Region of the Air. But seeing that Attraction hurried me up with so much rapidity, that instead of drawing near the Moon, as I intended, she seem'd to me to be more distant, than at my first setting out; I broke several of my Vials, until I found my weight exceed the force of the Attraction, and that I began to descend again towards the Earth. I was not mistaken in my opinion, for some time after I fell to the ground again; and to reckon from the hour that I set out at, it must then have been about midnight. Nevertheless I found the Sun to be in the Meridian, and that it was Noon. I leave it to you to judge, in what Amazement I was: The truth is, I was so strangely surprised, that not knowing what to think of that Miracle, I had the insolence to imagine, that in favour of my Boldness, God had once more nailed the Sun to the Firmament, to light so generous an Enterprise. That which encreased my Astonishment was, That I knew not the Country where I was; it seemed to me, that having mounted straight up, I should have fallen down again in the same place I parted from. However, in the Equipage I was in, I directed my course towards a kind of Cottage, where I perceived some smoke; and I was not above a Pistol-shot from it, when I saw my self environed by a great number of People, stark naked: They seemed to be exceedingly surprised at the sight of me; for I was the first, (as I think) that they had ever seen clad in Bottles. Nay, and to baffle all the Interpretations, that they could put upon that Equipage, they perceived, that I hardly touched the ground as I walked; for, indeed, they understood not, that upon the least agitation I gave my Body, the Heat of the beams of the Noon-Sun, raised me up with my Dew; and that if I had had Vials enough about me, it would possibly have carried me up

into

into the Air in their view. I had a mind to have spoken to them; but as if Fear had changed them into Birds, immediately I lost sight of them, in an adjoyning Forest. However, I catched hold of one, whose Legs had, without doubt, betrayed his Heart. I asked him, but with a great deal of pain, (for I was quite choked) how far they reckoned from thence to *Paris*? How long Men had gone naked in *France*? and why they fled from me in so great Consternation? The Man I spoke to was an old tawny Fellow, who presently fell at my Feet, and with lifted-up Hands, joyned behind his Head, opened his Mouth and shut his Eyes: He mumbled a long while between his Teeth, but I could not distinguish an articulate Word; so that I took his Language for the maffling noise of a Dumb-man.

Some time after, I saw a Company of Souldiers marching, with Drums beating; and I perceived Two detached from the rest, to come and take speech of me. When they were come within hearing, I asked them, Where I was? You are in *France*, answered they: But what Devil hath put you into that Dress? And how comes it that we know you not? Is the Fleet then arrived? Are you going to carry the News of it to the Governor?

And why have you divided your Brandy into so many Bottles? To all this I made answer, That the Devil had not put me into that Dress: That they knew me not; because they could not know all Men: That I knew nothing of the *Seine's* carrying Ships to *Paris*: That I had no news for the *Marshal de l' Hospital*; and that I was not loaded with Brandy. Ho, ho, said they to me, taking me by the Arm, you are a merry Fellow indeed; come, the Governor will make a shift to know you, no doubt on't. They led me to their Company, where I learnt that I was in reality in *France*, but that it was in *New-France*: So that some time after, I was presented before the Governor, who asked me my Country, my Name and Quality; and after that I had satisfied him in all Points, and told him the pleasant Success of my Voyage, whether he believed it, or only pretended to do so, he had the goodness to order me a Chamber in his Apartment. I was very happy, in meeting with a Man capable of lofty Opinions, and who was not at all surprised, when I told him, that the Earth must needs have turned during my Elevation; seeing that having begun to mount about Two Leagues from *Paris*, I was fallen, as it were, by a perpendicular Line in *Canada*.

When

When I was going to Bed at night, he came into my Chamber, and spoke to me to this purpose: I should not have come to disturb your Rest, had I not thought that one who hath found out the secret, of Travelling so far in Twelve hours space, had likewise a charm against Lassitude. But you know not, added he, what a pleasant Quarrel I have just now had with our Fathers, upon your account? They'll have you absolutely to be a Magician; and the greatest favour you can expect from them, is to be reckoned only an Impostor: The truth is, that Motion which you attribute to the Earth, is a pretty nice Paradox; and for my part I'll frankly tell you, That that which hinders me from being of your Opinion, is, That though you parted yesterday from *Paris*, yet you might have arrived to day in this Country, without the Earth's turning: For the Sun having drawn you up, by the means of your Bottles, ought he not to have brought you hither; since according to *Ptolemy*, and the Modern Philosophers, he marches obliquely, as you make the Earth to move? And besides, what great Probability have you to imagine, that the Sun is immoveable, when we see it go? And what appearance is there, that the Earth turns with so great Rapidity, when we feel it firm under our Feet? Sir, replied

replied I to him, These are, in a manner, the Reasons that oblige us to think so: In the first place, it is consonant to common Sense, to think that the Sun is placed in the Center of the Universe; seeing all Bodies in nature, standing in need of that radical Heat, it is fit he should reside in the heart of the Kingdom, that he may be in a condition, readily to supply the Necessities of every Part; and that the Cause of Generations, should be placed in the middle of all Bodies, that it may act there with greater Equality and Ease: After the same manner, as Wise Nature hath placed the Genitals in Man, the Seeds in the Center of Apples, the Kernels in the middle of their Fruits; and in the same manner, as the Onion, under the cover of so many Coats that encompass it, preserves that precious Bud, from which Millions of others are to have their being, for an Apple is in it self a little Universe; the Seed, hotter than the other parts thereof, is its Sun, which diffuses about it self that natural Heat, which preserves its Globe: And in the Onion, the Germ is the little Sun of that little World, which vivifies and nourishes the vegetative Salt of that little mass. Having laid down this, then, for a ground, I say, That the Earth standing in need of the Light, Heat, and Influence of this great Fire, it turns round it, that it

may

may receive in all parts alike, that Virtue which keeps it in Being. For it would be as ridiculous to think, that that vast luminous Body, turned about a point, that it has not the least need of; as to imagine, that when we see a roasted Lark, that the Kitchin-fire must have turned round it. Else, were it the part of the Sun to do that drudgery, it would seem that the Physician stood in need of the Patient; that the Strong should yield to the Weak; the Superior serve the Inferior; and that the Ship did not sail about the Land, but the Land about the Ship. Now if you cannot easily conceive, how so ponderous a Body can move; Pray, tell me, are the Stars and Heavens, which, in your Opinion, are so solid, any way lighter? Besides, it is not so difficult for us, who are assured of the Roundness of the Earth, to infer its motion from its Figure: But why do ye suppose the Heaven to be round, seeing you cannot know it, and that yet, if it hath not this Figure, it is impossible it can move? I object not to you your *Excentricks* nor *Epicycles*, which you cannot explain but very confusedly, and which are out of doors in my Systeme. Let's reflect only on the natural Causes of that Motion. To make good your Hypothesis, you are forced to have recourse to Spirits or *Intelligences*, that move and govern

vern your Spheres. But for my part, without disturbing the repose of the supreme Being, who, without doubt, hath made Nature entirely perfect, and whose Wisdom ought so to have compleated her, that being perfect in one thing, she should not have been defective in another: I say, that the Beams and Influences of the Sun, darting Circularly upon the Earth, make it to turn, as with a turn of the Hand, we make a Globe to move; or, which is much the same, that the Steams which continually evaporate from that side of it, which the Sun shines upon, being reverberated by the Cold of the middle Region, rebound upon it, and striking obliquely, do of necessity make it whirle about in that manner.

The Explication of the other Motions is less perplexed still; for pray, consider a little----- At these words the Vice-Roy interrupted me: I had rather, said he, you would excuse your self from that trouble; for I have read some Books of *Gassendus* on that subject: And hear what one of our Fathers, who maintained your Opinion one day, answered me. Really, said he, I fancy that the Earth does move, not for the Reasons alledged by *Copernicus*; but because Hell-fire, being shut up in the Center of the Earth, the damned who make a

great

World of the Moon.

great bustle to avoid its Flames, scramble up to the Vault, as far as they can from them, and so make the Earth to turn, as a Turn-pit makes the Wheel go round, when he [r]uns about in it.

We applauded that Thought, as being [a] pure effect of the Zeal of that good Fa[t]her: And then the Vice-Roy told me, [t]hat he much wondered, how the Systeme [o]f *Ptolemy*, being so improbable, should [h]ave been so universally received. Sir, said [I] to him, most part of Men, who judge of [a]ll things by the Senses, have suffered them[se]lves to be perswaded by their Eyes; and [a]s he who Sails along a Shoar, thinks the [S]hip immoveable, and the Land in motion; [ev]en so Men turning with the Earth round [th]e Sun, have thought that it was the Sun [th]at moved about them. To this may be [ad]ded, the unsupportable Pride of Mankind, [w]ho perswade themselves, that Nature [ha]th only been made for them; as if it were [lik]ely that the Sun, a vast Body, Four [hu]ndred and thirty four times bigger than [th]e Earth, had only been kindled to ripen [th]eir Medlars, and plumpen their Cabbage. [Fo]r my part, I am so far from complying [wi]th their Insolence, that I believe the Pla[ne]ts are Worlds about the Sun, and that [the] fixed Stars are also Suns, which have [Plan]ets about them, that's to say, Worlds,

which

which because of their smallness, and that their borrowed light cannot reach us, are not discernable by Men in this World: For in good earnest, how can it be imagined, that such spacious Globes are no more but vast Desarts; and that ours, because we live in it, hath been framed for the habitation of a dozen of proud Dandyprats? How, must it be said, because the Sun measures our Days and Years, that it hath only been made, to keep us from running our Heads against the Walls? No, no, if that visible Deity shine upon Man, it's by accident, as the King's Flamboy by accident lightens a Porter that walks along the Street: But, said he to me, if, as you affirm, the fixed Stars be so many Suns, it will follow, that the World is infinite; seeing it is probable, that the People of that World, which moves about that fixed Star, you take for a Sun, discover above themselves other fixed Stars, which we cannot perceive from hence, and so others in that manner *in infinitum*.

Never question, replied I, but as God could create the Soul Immortal, He could also make the World Infinite; if so it be that Eternity is nothing else, but an illimited Duration, and an *infinite*, a boundless Extension: And then God himself would be Finite, supposing the World not to be

infi

infinite, seeing he cannot be where nothing is, and that he could not encrease the greatness of the World, without adding somewhat to his own Being, by beginning to exist, where he did not exist before. We must believe then, that as from hence we see *Saturn* and *Jupiter*; if we were in either of the Two, we should discover a great many Worlds which we perceive not, and that the Universe extends so *in infinitum.*

I' faith, replied he, when you have said all you can, I cannot at all comprehend that Infinitude. Good now, replied to him, do you comprehend the Nothing that is beyond it? Not at all. For when you think of that *Nothing*, you imagine it at least to be like Wind or Air, and that is a Being: But if you conceive not an *Infinite* in general, you comprehend it at least in particulars; seeing it is not difficult to fancy to our selves beyond the Earth, Air, and Fire which we see, other Air, and other Earth, and other Fire. Now Infinitude is nothing else, but a boundless Series of all these. But if you ask me, How those Worlds have been made, seeing Holy Scripture speaks only of one that God made? My answer is, That I have no more to say; for to oblige me to give a Reason for every thing, that comes into my Imagination, is to to stop my Mouth, and make me confess, that in things of that nature, my Reason shall always stoop to Faith. He ingeniously acknowledged to me, that his Question was to be censured, but bid me pursue my notion: So that I went on, and told him, That all the other Worlds, which are not seen, or but imperfectly believed, are no more, but the Scum that purges out of the Suns. For how could these great Fires subsist without some matter, that served them for Fewel? Now as the Fire drives from it the Ashes that would stifle it, or the Gold in a Crucible, separates from the Marcasite and Dross, and is refined to the highest Standard; nay, and as our Stomack discharges it self by vomit, of the Crudities that oppress it; even so these Suns daily evacuate, and reject the Remains of matter, that might incommode their Fire: But when they have wholly consumed that matter, which entertains them; you are not to doubt, but they spread themselves abroad on all sides, to seek for fresh Fewel, and fasten upon the Worlds, which heretofore they have made, and particularly upon those that are nearest: Then these great Fires, reconcocting all the Bodies, will as formerly force them out again, *Pell-mell* from all parts; and being by little and little purified, they'll begin to serve for

Suns

Suns, to other little Worlds, which they procreate by driving them out of their Spheres: And that without doubt, made the *Pythagoreans* foretel the univerfal Conflagration.

This is no ridiculous Imagination, for *New-France* where we are, gives us a very convincing inftance of it. The vaft Continent of *America*, is one half of the Earth, which in fpight of our Predeceffors, who a Thoufand times had cruifed the Ocean, was not at that time difcovered: Nor, indeed, was it then in being, no more than a great many Iflands, Peninfules, and Mountains that have fince ftarted up in our Globe; when the Sun purged out its Excrements to a convenient diftance, and fufficient Gravity, to be attracted by the Center of our World, either in fmall Particles, perhaps, or, it may be alfo, altogether in one lump. That is not fo unreafonable, but that St. *Auftin* would have applauded to it, if that Country had been difcovered in his Age. Seeing that great Man, who had a very clear Wit, affures us, That in his time, the Earth was flat like the floor of an Oven, and that it floated upon the Water, like the half of an Orange: But if ever I have the honour to fee you in *France*, I'll make you obferve, by means of a moft excellent Celefcope, that fome Obfcurities, which from hence appear to be Spots, are Worlds a forming.

My Eyes that shut with this Discourse, obliged the Vice-Roy to withdraw. Next Day, and the Days following, we had some Discourses to the same purpose: But some time after, since the hurry of Affairs suspended our Philosophy, I fell afresh upon the design of mounting up to the Moon.

So soon as she was up, I walked about musing in the Woods, how I might manage and succeed in my Enterprise; and at length on St. *John*'s-Eve, when they were at Council in the Fort, whether they should assist the Wild Natives of the Country against the *Iroqueans*; I went all alone to the top of a little Hill, at the back of our Habitation, where I put in Practice what you shall hear. I had made a Machine, which I fancied might carry me up as high as I pleased, so that nothing seeming to be wanting to it, I placed my self within, and from the Top of a Rock, threw my self in the Air: But because I had not taken my measures aright, I fell with a sosh in the Valley below. Bruised as I was, however, I returned to my Chamber, without loosing courage, and with Beef-Marrow I anointed my Body, for I was all over mortified from Head to Foot: Then having taken a

dram

dram of Cordial Waters to strengthen my Heart, I went back to look for my Machine; but I could not find it, for some Soldiers, that had been sent into the Forest, to cut wood for a Bonefire, meeting with it by chance, had carried it with them to the Fort: Where after a great deal of guessing what it might be, when they had discovered the invention of the Spring, some said, that a good many Fire-Works should be fastened to it, because their Force carrying them up on high, and the Machine playing its large Wings, no Body but would take it for a Fiery Dragon. In the mean time I was long in search of it, but found it at length in the middle of the Market-place of *Kebeck*, just as they were setting Fire to it. I was so transported with Grief, to find the Work of my Hands in so great Peril, that I ran to the Souldier, that was giving Fire to it, caught hold of his Arm, pluckt the Match out of his Hand, and in great rage threw my self into my Machine, that I might undo the Fire-Works, that they had stuck about it; but I came too late, for hardly were both my Feet within, when whip, away went I up in a Cloud: The Horror and Consternation I was in, did not so confound the faculties of my Soul, but I have since remembred all that happened to me at that instant. For so soon as the Flame had devoured one tire of Squibs, which were ranked by six and six, by means of a Train, that reached every half-dozen; another tire went off, and then another; so that the Salt-Peter taking Fire, put off the danger by encreasing it. However, all the combustible matter being spent, there was a period put to the Fire-work; and whilst I thought of nothing less, than to knock my Head against the top of some Mountain, I felt, without the least stirring, my elevation continuing; and adieu Machine, for I saw it fall down again towards the Earth. That extraordinary Adventure puffed up my Heart with so uncommon a Gladness; that, ravished to see my self delivered from certain danger, I had the impudence to philosophize upon it. Whilst then with Eyes and Thought I cast about, to find what might be the cause of it, I perceived my flesh blown up, and still greasy with the Marrow, that I had daubed my self over with, for the Bruises of my fall: I knew that the Moon being then in the Wain, and that it being usual for her in that Quarter, to suck up the Marrow of Animals; she drank up that wherewith I was anointed, with so much the more force, that her Globe was nearer to me, and that no interposition of Clouds weakened her Attraction.

When

World of the Moon.

When I had, according to the computation I made since, advanced a good deal more, than three quarters of the space, that divided the Earth from the Moon; all of a sudden I fell with my Heels up, and Head down, though I had made no Trip; and indeed, I had not been sensible of it, had not I felt my Head loaded under the weight of my Body: The truth is, I knew very well, that I was not falling again towards our World; for though I found myself to be betwixt two Moons, and easily observed, that the nearer I drew to the one, the farther I removed from the other; yet I was certain, that ours was the bigger Globe of the two: Because after one or two days Journey, the remote Refractions of the Sun, confounding the diversity of Bodies and Climates, it appeared to me only as a large Plate of Gold: That made me imagine, that I byassed towards the Moon; and I was confirmed in that opinion, when I began to call to mind, that I did not fall, till I was past three quarters of the way. For, said I to myself, that Mass being less than ours, the Sphere of its Activity must be of less Extent also; and by consequence, it was later before I felt the force of its Center.

In fine, after I had been a very long while in falling, as I judged, for the violence

olence of my Precipitation hindered me from obferving it more exactly: The laſt thing I can remember is, That I found my ſelf under a Tree, entangled with three or four pretty large Branches, which I had broken off by my fall; and my Face beſmeared with an Apple, that had daſhed againſt it.

By good luck that place was, as you ſhall know by and by -------- So that you may very well conclude, that had it not been for that Chance, if I had had a thouſand lives, they had been all loſt. I have many times ſince reflected upon the vulgar Opinion, That if one precipitate himſelf from a very high place, his breath is out before he reach the ground; and from my adventure I conclude it to be falſe, or elſe that the efficacious Juyce of that Fruit, which ſquirted into my mouth, muſt needs have recalled my Soul, that was not far from my Carcaſs, which was ſtill hot, and in a diſpoſition of exerting the Functions of Life. The truth is, ſo ſoon as I was upon the ground, my pain was gone, before I could think what it was; and the Hunger, which I felt during my Voyage, was fully ſatisfied with the ſenſe that I had loſt it.

When I was got up, I had hardly taken notice of the largeſt of Four great Rivers,

Rivers, which by their conflux make a Lake; when the Spirit, or invisible Soul of Plants, that breath upon that Country, refreshed my Brain with a delightful smell: And I found that the Stones there, were neither hard nor rough; but that they carefully softened themselves, when one trode upon them. I presently lighted upon a Walk with five Avenues, in figure like to a Star; the Trees whereof seemed to reach up to the Skie, a green plot of lofty Boughs: Casting up my Eyes from the root to the top, and then making the same Survey downwards, I was in doubt whether the Earth carried them, or they the Earth, hanging by their Roots: Their high and stately Forehead seemed also to bend, as it were by force, under the weight of the Celestial Globes; and one would say, that their Sighs and out-stretched Arms, wherewith they embraced the Firmament, demanded of the Stars the bounty of their purer Influences, before they had lost any thing of their Innocence, in the contagious Bed of the Elements. The Flowers there on all hands, without the aid of any other Gardiner but Nature, send out so sweet (though wild) a Perfume, that it rouzes and delights the Smell: There the incarnate of a Rose upon the Bush, and the lively Azure of a

Violet under the Rushes, captivating the Choice, make each of themselves to be judged the Fairest: There the whole Year is a Spring; there no poysonous Plant sprouts forth, but is as soon destroyed; there the Brooks by an agreeable murmuring, relate their Travels to the Pebbles; there Thousands of Quiristers make the Woods, resound with their melodious Notes; and the quavering Clubs of these divine Musicians are so universal, that every Leaf of the Forest, seems to have borrowed the Tongue and shape of a Nightingale; nay, and the Nymph *Eccho* is so delightful with their Airs, that to hear her repeat, one would say, She were sollicitous to learn them. On the sides of that Wood, are Two Meadows, whose continued Verdure seems an Emerauld, reaching out of sight. The various Colours, which the Spring bestows upon the numerous little Flowers that grow there, so delightfully confounds and mingles their Shadows; that it is hard to be known, whether these Flowers shaken with a gentle Breeze, pursue themselves, or fly rather from the Caresses of the Wanton *Zephyrus*; one would likewise take that Meadow for an Ocean, because as the Sea, it presents no Shoar to the view; insomuch, that mine Eye fearing it might lose it self, having roamed so long, and disco-

discovered no Coast, sent my Thoughts presently thither; and my Thoughts, imagining it to be the end of the World, were willing to be perswaded, that such charming places, had, perhaps, forced the Heavens to descend, and join the Earth there. In the midst of that vast and pleasant Carpet, a rustick Fountain bubbles up in Silver Purles, crowning its enamelled Banks with Sets of Violets, and multitudes of other little Flowers, that seem to strive, which shall first behold it self in that Chrystal Myrroir: It is as yet in the Cradle, being but newly Born, and its Young and smooth Face shews not the least Wrinkle. The large Compasses it fetches, in circling within it self, demonstrate its unwillingness to leave its native Soyl: And as if it had been ashamed to be caressed in presence of its Mother, with a Murmuring it thrust back my hand, that would have touched it: The Beasts that came to drink there, more rational than those of our World, seemed surprised to see it day upon the Horizon, whilst the Sun was with the *Antipodes*; and durst not bend downwards upon the Brink, for fear of falling into the Firmament.

 I must confess to you, That at the sight of so many Fine things, I found my self tickled with these agreeable Twitches,

which,

which, they say, the *Embryo* feels upon the infusion of its Soul: My old Hair fell off, and gave place for thicker and softer Locks: I perceived my Youth revived, my Face grow ruddy, my natural Heat mingle gently again, with my radical Moisture: And in a word, I grew younger again, by at least Fourteen Years.

I had advanced half a League, through a a Forest of Jessamines and Myrtles, when I perceived something that stirred, lying in the Shade: It was a Youth, whose Majestick Beauty forced me almost to Adoration. He started up to hinder me; crying, It is not to me, but to God that you owe these Humilities. You see one, answered I, stunned with so many Wonders, that I knew not what to admire most; for coming from a World, which without doubt you take for a Moon here, I thought I had arrived in another, which our Worldlings call a Moon also; and behold I am in Paradice at the Feet of a God, who will not be Adored. Except the quality of a God, replied he, whose Creature I only am, the rest you say is true: This Land is the Moon, which you see from your Globe, and this place where you are is ——— Now at that time Man's Imagination was so strong, as not being as yet corrupted, neither by Debauches, the Crudity of Aliments,

ments, nor the alterations of Diseases, that being excited by a violent desire of coming to this Sanctuary, and his Body becoming light, through the heat of this Inspiration; he was carried thither in the same manner, as some Philosophers, who having fixed their Imagination, upon the contemplation of a certain Object, have sprung up in the Air by Ravishments, which you call Extasies. The Woman, who through the infirmity of her Sex, was weaker and less hot, could not, without doubt, have the Imagination strong enough, to make the Intension of her Will, prevail over the Ponderousness of her Matter; but because there were very few ———

The Sympathy which still united that half to its whole, drew her towards him as he mounted up, as the Amber attracts the Straw; the Load-stone turns towards the North, from whence it hath been taken, and drew to him that part of himself, as the Sea draws the Rivers which proceed from it. When they arrived in your Earth, they dwelt betwixt *Mesopotamia* and *Arabia*: Some People knew them by the name of ———, and others under that of *Prometheus*, whom the Poets feigned to have stolen Fire from Heaven, by reason of his Off-spring, who were endowed with a Soul as perfect as his own: So that

to

to inhabit your World, that Man left this destitute; but the All-wise, would not have so blessed an Habitation, to remain without Inhabitants: He suffered a few ages after that ------- cloyed with the company of Men, whose Innocence was corrupted, had a desire to forsake them. This person, however, thought no retreat secure enough from the Ambition of Men, who already Murdered one another, about the distribution of your World; except that blessed Land, which his Grand-Father had so often mentioned unto him, and to which no Body had as yet found out the way: But his Imagination supplied that; for seeing he had observed that ------- he filled Two large Vessels which he sealed Hermetically, and fastened them under his Arm-pits: So soon as the Smoak began to rise upwards, and could not pierce through the Mettal, it forced up the Vessels on high, and with them also that Great Man. When he was got as high as the Moon, and had cast his Eyes upon that lovely Garden, a fit of almost supernatural Joy convinced him, that that was the place, where his Grandfather had heretofore lived. He quickly untied the Vessels, which he had girt like Wings about his Shoulders, and did it so luckily, that he was scarcely Four Fathom in the Air above the Moon, when he set
his

his Fins a going; yet he was high enough still, to have been hurt by the fall, had it not been for the large skirts of his Gown, which being swelled by the Wind, gently upheld him, till he set Foot on ground. As for the two Vessels, they mounted up to a certain place, where they have continued: And those are they, which now a-days you call the *Balance*.

I must now tell you, the manner how I came hither: I believe you have not forgot my name, seeing it is not long since I told it you. You shall know then, that I lived on the agreeable Banks of one of the most renowned Rivers of your World; where amongst my Books, I lead a Life pleasant enough, not to be lamented, though it slipt away fast enough. In the mean while, the more I encreased in Knowledge, the more I knew my Ignorance. Our Learned Men never put me in mind of the famous *Mada*, but the thoughts of his perfect Philosophy made me to Sigh. I was despairing of being able to attain to it, when one day, after a long and profound Studying, I took a piece of Load-stone about two Foot square, which I put into a Furnace; and then after it was well purged, precipitated and dissolved, I drew the calcined Attractive of it, and reduced it into the size of about an ordinary Bowl.

Aft

After these Preparations, I got a very light Machine of Iron made, into which I went ------- and when I was well seated in my place, I threw this Magnetick Bowl, as high as I could, up into the Air. Now the Iron Machine, which I had purposely made more massive in the middle than at the ends, was presently elevated, and in a just Poise; because the middle received the greatest force of Attraction. So then, as I arrived at the place, whither my Loadstone had attracted me, I presently threw up my Bowl in the Air over me. But, said I, interrupting him, How came you to heave up your Bowl so streight over your Chariot, that it never happened to be on one side of it? That seems to me to be no wonder at all, said he; for the Load-stone being once thrown up in the Air, drew the Iron streight towards it; and so it was impossible, that ever I should mount side-ways. Nay more, I can tell you, that when I held the Bowl in my hand, I was still mounting upwards; because the Chariot flew always to the Load-stone, which I held over it. But the effort of the Iron, to be united to my Bowl, was so violent, that it made my Body bend double; so that I durst but once essay that new Experiment. The truth is, it was a very surprizing Spectacle to behold; for the Steel

Steel of that flying House, which I had very carefully Polished, reflected on all sides the light of the Sun, with so great life and lustre, that I thought my self to be all on fire. In fine, after often Bowling and following of my Cast, I came, as you did, to an Elevation, from which I descended towards this World; and because at that instant, I held my Bowl very fast between my hands, my Machine, whereof the Seat pressed me hard, that it might approach its Atractive, did not forsake me; all that now I feared was, that I should break my Neck: But to save me from that, ever now and then, I tossed up my Bowl; that by its atractive Virtue, it might prevent the violent Descent of my Machine, and render my fall more easie, as indeed it happened; for when I saw my self, within Two or three hundred fathom of the Earth, I threw out my Bowl on all hands, level with the Chariot, sometimes on this side, and sometimes on that, until I came to a certain Distance; and immediately then, I tossed it up above me; so that my Machine following it, I left it, and let my self fall on the other side, as gently as I could, upon the Sand; insomuch that my fall was no greater, than if it had been but my own height. I shall not describe to you

you the amazement I was in at the sight of the wonders of this place, seeing it was so like the same, wherewith I just now saw you seized ⸺

Scarcely had I tasted it, when a thick Cloud over-cast my Soul: I saw no body now near me; and in the whole Hemisphere, my Eyes could not discern the least Tract of the way I had made; yet, nevertheless I fully remembred every thing that befel me. When I reflected since upon that Miracle, I fanced that the skin of the Fruit which I bit, had not rendered me altogether brutish; because my Teeth piercing through it were, a little moistened by the Juyce within, the efficacy whereof had dissipated the Malignities of the Rind. I was not a little surprised to see my self all alone, in a Country I knew not. It was to no purpose for me, to stare and look about me; for no Creature appeared to comfort me: At length, I resolved to march forwards, till Fortune should aford me the company of some Beasts, or at least the means of Dying.

She favourably granted my desire; for within half a quarter of a League, I met two huge Animals, one of which stopt before me, and the other fled swiftly to its Den; for so I thought at least; because that some time after, I perceived it come back again,

in

in company of above Seven or Eight hundred of the same kind, who beset me. When I could discern them at a near distance, I perceived that they were proportioned and shaped like us. This adventure brought into my mind, the old Wives Tales of my Nurse concerning *Syrenes*, *Faunes* and *Satyrs*: Ever now and then they raised such furious Shouts, occasioned undoubtedly by their Admiration, at the sight of me, that I thought I was e'en turned a Monster. At length one of these Beast-like men, catching hold of me by the Neck, just as Wolves do when they carry away Sheep, tossed me upon his back, and brought me into their Town; where I was more amazed than before, when I knew they were Men, that I could meet with none of them, but who marched upon all four.

When these People saw that I was so little, (for most of them are Twelve Cubits long,) and that I walked only upon Two Legs, they could not believe me to be a Man: For they were of opinion, that Nature having given to men as well as Beasts Two Legs and Two Arms, they should both make use of them alike. And, indeed, reflecting upon that since, that scituation of Body did not seem to me altogether extravagant; when I called to mind, that whilst Children are
still

still under the nurture of Nature, they go upon all four, and that they rise not on their two Legs, but by the care of their Nurses; who set them in little running Chairs, and fasten straps to them, to hinder them from falling on all four, as the only posture that the shape of our Body naturally inclines to rest in.

They said then, (as I had it interpreted to me since) That I was infallibly the Female of the Queens little Animal. And therefore as such, or somewhat else, I was carried streight to the Town-House, where I observed by the muttering and gestures, both of the People and Magistrates, that they were consulting what sort of a thing I could be. When they had conferred together a long while, a certain Burgher, who had the keeping of the strange Beasts, besought the Mayor and Aldermen to commit me to his Custody, till the Queen should send for me to couple me to my Male. This was granted without any difficulty, and that Juggler carried me to his House; where he taught me to Tumble, Vault, make Mouths, and shew a Hundred odd Tricks, for which in the Afternoons he received Money at the door, from those that came in to see me: But Heaven pitying my Sorrows, and vext to see the Temple of its Maker profaned, so

ordered

ordered it, that one day as I was tied to a Rope, wherewith the Mountebank made me Leap and Skip to divert the People; I heard a Man's voice, who asked me what I was, in Greek. I was much furprifed to hear one fpeak in that Country, as they do in our World. He put fome Queftions to me, which I anfwered, and then gave him a full account of my whole defign, and the fuccefs of my Travels: He took the pains to comfort me, and, as I take it, faid to me: Well, Son, at length you fuffer for the frailties of your World: There is a Mobile here, as well as there, that can away with nothing, but what they are accuftomed to: But know, that you are but juftly ferved; for had any one of this Earth, had the boldnefs to mount up to yours, and call himfelf a Man, your Sages would have deftroyed him as a Monfter. He then told me, That he would acquaint the Court with my difafter; adding, that fo foon as he had heard the news that went of me, he came to fee me, and was fatisfied that I was a man of the World, of which I faid I was; becaufe he had Travelled there formerly, and fojourned in *Greece*, where he was called the *Demon* of *Socrates*: That after the Death of that Philofopher, he had governed and taught *Epaminondas* at *Thebes*: After which being gone over to the *Romans*, Juftice had

had obliged him to espouse the party of the Younger *Cato*: That after his Death, he had addicted himself to *Brutus*: That all these great Men, having left in that World no more, but the shadow of their Virtues, he with his Companions, had retreated to Temples and Solitudes. In a word, added he, the People of your World became so dull and stupid, that my Companions and I lost all the Pleasure, that formerly we had had in instructing them: Not but that you have heard Men talk of us; for they called us *Oracles, Nymphs, Geniuses, Fairies, Houshold-Gods, Lemmes, Larves, Lamiers, Hobgoblins, Nayades, Incubusses, Shades, Manes, Visions* and *Apparitions*: We abandoned your World, in the Reign of *Augustus*, not long after I had appeared to *Drusus* the Son of *Livia*, who waged War in *Germany*, whom I forbid to proceed any farther. It is not long since I came from thence a second time; within these Hundred Years I had a Commission to Travel thither: I roamed a great deal in *Europe*, and conversed with some, whom possibly you may have known. One Day, amongst others, I appeared to *Cardan*, as he was at his Study; I taught him a great many things, and he in acknowledgment promised me, to inform Posterity, of whom he had those Wonders, which

which he intended to leave in writing. There I saw *Agrippa*, the Abbot *Trithemius*, Doctor *Faustus*, *La Brosse*, *Cæsar*, and a certain Cabal of Young Men, who are commonly called *Rosacrucians* or *Knights of the Red-Cross*, whom I taught a great many Knacks and Secrets of Nature, which, without doubt, have made them pass for great Magicians: I knew *Campanella* also; it was I that advised him, whilst he was in the Inquisition at *Rome*, to put his Face and Body into the usual Postures of those, whose inside he needed to know, that by the same frame of Body, he might excite in himself, the thoughts which the same scituation had raised in his Adversaries; because by so doing, he might better manage their Soul, when he came to know it; and at my desire he began a Book, which we Entituled, *De Sensu Rerum*. I likewise haunted in *France*, *La Mothe le Vayer* and *Gassendus*; this last hath written as much like a Philosopher, as the other lived: I have known a great many more there, whom our Age call *Divines*, but all that I could find in them, was a great deal of Babble, and a great deal of Pride. In fine, since I past over from your Country into *England*, to acquaint my self, with the maners of its Inhabitants, I met with a Man, the shame of his Country; for certainly it

is

is a great shame for the Grandees of your States, to know the virtue which in him has its Throne, and not to adore him: That I may give you an Abridgement of his Panegyrick, he is all Wit, all Heart, and possesses all the Qualities, of which one alone was heretofore sufficient to make an Heroe: It was *Tristan* the Hermite. The Truth is, I must tell you, when I perceived so exalted a Virtue, I mistrusted it would not be taken notice of, and therefore I endeavoured to make him accept Three Vials, the first filled with the Oyl of Talk, the other with the Powder of Projection, and the third with *Aurum Potabile*; but he refused them with a more generous Disdain, than *Diogenes* did the Complements of *Alexander*. In fine, I can add nothing to the Elogy of that Great Man, but that he is the only Poet, the only Philosopher, and the only Free-man amongst you: These are the considerable Persons that I conversed with; all the rest, at least, that I know, are so far below Men, that I have seen Beasts somewhat above them.

After all, I am not a Native neither of this Country nor yours, I was born in the Sun; but because sometimes our World is over-stock'd with people, by reason of the long Lives of the Inhabitants, and that

there

here is hardly any Wars, or Diseases amongst them: Our Magistrates, from time to time, send Colonies into the neigbouring Worlds. For my own part, I was commanded to go to yours; being declared Chief of the Colony that accompanyed me.

I came since into this World, for the Reasons I told you; and that which makes me continue here, is, because the Men are great lovers of Truth, have no Pedants among them; that the Philosophers are never perswaded but by Reason, and that the Authority of a Doctor, or of a great number, is not preferred before the Opinion of a Thresher in a Barn, when he has right on his side. In short, none are reckoned Mad-men in this Country, but Sophisters and Orators. I asked him how they lived? he made answer, three or four thousand Years; and thus went on:

Though the Inhabitants of the Sun, be not so numerous as those of this World; yet the Sun is many times over stocked, because the People being of a hot constitution, are stirring and ambitious, and digest much.

You ought not to be surprised at what I tell you; for though our Globe be very vast, and yours little, though we die not before the end of Four thousand Years, and you at the end of Fifty; yet know, that as there are

are not so many Stones as clods of Earth, nor so many Animals as Plants, nor so many Men as Beasts; just so there ought not to be so many Spirits as Men, by reason of the difficulties that occur in the Generation of a perfect Creature.

I asked him, if they were Bodies as we are? He made answer, That they were Bodies, but not like us, nor any thing else which we judged such; because we call nothing a Body commonly, but what we can touch: That, in short, there was nothing in Nature, but what was material; and that though they themselves were so, yet they were forced, when they had a mind to appear to us, to take Bodies proportionated to what our Senses are able to know; and that, without doubt, that was the reason, why many have taken the Stories that are told of them, for the Delusions of a weak Fancy, because they only appeared in the night time: He told me withal, That seeing they were necessitated to piece together the Bodies, they were to make use of in great haste, many times they had not leisure enough, to render them the Objects of more Senses than one at a time, sometimes of the Hearing, as the Voices of *Oracles*, sometimes of the Sight, as the *Fires* and *Visions*, sometimes of the Feeling, as the *Incubusses*;
and

and that thefe Bodies being but Air condenfed, in fuch or fuch a manner, the Light difperfed them by its heat, in the fame manner, as it fcatters a Mift.

So many fine things as he told me, gave me the curiofity to queftion him about his Birth and Death; if in the Country of the Sun, the *individual* was procreated by the ways of Generation, and if it died by the diffolution of its Conftitution, or the difcompofure of its Organs? Your fenfes, replied he, bear but too little proportion to the Explication of thefe Myfteries: Ye Gentlemen imagine, that whatfoever you cannot comprehend is fpiritual, or that it is not at all; but that Confequence is abfurd, and it is an argument, that there are a Million of things, perhaps, in the Univerfe, that would require a Million of different Organs in you, to underftand them. For inftance, I by my Senfes know the caufe of the Sympathy, that is betwixt the Loadftone and the Pole, of the ebbing and flowing of the Sea, and what becomes of the Animal after Death; you cannot reach thefe high Conceptions but by Faith, becaufe they are Secrets above the power of your Intellects; no more than a Blind-man can judge of the beauties of a Land-skip, the Colours of a Picture, or the ftreaks of a Rain-bow; or at beft he will fancy them to besomewhat palpable, to be like Eating, a Sound, or a pleasant Smell: Even so, should I attempt to explain to you, what I perceive by the Senses which you want, you would represent it to your self, as somewhat that may be Heard, Seen, Felt, Smelt or Tasted, and yet it is no such thing.

He was gone on so far in his Discourse, when my Juggler perceived, that the Company began to be weary of my Gibberish, that they understood not, and which they took to be an inarticulated Grunting: He therefore fell to pulling my Rope afresh, to make me leap and skip, till the Spectators having had their Belly-fulls of Laughing, affirmed that I had almost as much Wit, as the Beasts of their Country, and so broke up.

Thus, all the comfort I had during the misery of my hard Usage, were the visits of this officious Spirit; for you may judge what conversation I could have, with these that came to see me, since besides that they only took me for an Animal, in the highest class of the *Category* of Bruits, I neither understood their Language, nor they mine. For you must know, that there are but two Idioms in use in that Country, one for the Grandees, and another for the People in general.

That

That of the great ones is no more, but [v]arious inarticulate Tones, much like [t]o our Musick, when the Words are not [a]dded to the Air: and in reality it is an [In]vention, both very useful and pleasant; [fo]r when they are weary of talking, or dis[d]ain to prostitute their Throats to that Of[fi]ce, they take either a Lute, or some other [I]nstrument, whereby they communicate [th]eir Thoughts, as well as by their Tongue: [S]o that sometimes Fifteen or Twenty in a [C]ompany, will handle a point of Divinity, [o]r discuss the difficulties of a Law-suit, in [th]e most harmonious Consort, that ever [ti]ckled the Ear.

The second, which is used by the Vulgar, [is] performed by a shivering of the Mem[b]ers, but not, perhaps, as you may ima[g]ine, for some parts of the Body sig[n]ifie an entire Discourse; for example, the [a]gitation of a Finger, a Hand, an Ear, [a] Lip, an Arm, an Eye, a Cheek, every [o]ne severally will make up an Oration, or a [P]eriod with all the parts of it: Others [se]rve only instead of Words, as the knit[ti]ng of the Brows, the several quiverings [o]f the Muscles, the turning of the Hands, [th]e stamping of the Feet, the contorsion [o]f the Arm; so that when they speak, as [th]eir Custom is, stark naked, their Mem[b]ers being used to gesticulate their Conceptions,

tions, move so quick, that one would not think it to be a Man that spoke, but a Body that trembled.

Every day almost the Spirit came to see me, and his rare Conversation made me patiently bear with the rigour of my Captivity. At length, one morning I saw a Man enter my Cabbin, whom I knew not, who having a long while licked me gently, took me up in his Teeth by the Shoulder, and with one of his Paws, wherewith he held me up, for fear I might hurt my self, threw me upon his Back; where I found my self so softly seated, and so much at my ease, that being afflicted to be used like a Beast, I had not the least desire of making my escape; and besides, these Men that go upon all four, are much swifter than we, seeing the heaviest of them, make nothing of running down a Stagg.

In the mean time I was extreamly troubled, that I had no news of my courteous Spirit; and the first night we came to our Inn, as I was walking in the Court, expecting till Supper should be ready, a pretty handsome young Man came smiling in my Face, and cast his Two Fore-Legs about my Neck. After I had a little considered him : How! said he in *French*, do not you know your Friend then ? I leave you to judge in what
case

Powder and Shot, which Kills, Plucks, Roasts, and Seasons the Fowl all at once. I took up some of them, and eat them upon his word; and to say the Truth, In all my Life time, I never eat any thing so delicious. Having thus Breakfasted, we prepared to be gone, and with a Thousand odd Faces, which they use when they would shew their Love; our Landlord received a Paper from my Spirit. I asked him, if it was a Note for the Reckoning? He replied, No, that all was paid, and that it was a Copy of Verses. How! Verses, said I, are your Inn-Keepers here curious of Rhime then? It's, said he, the Money of the Country, and the charge we have been at here, hath been computed to amount to Three *Couplets*, or Six Verses, which I have given him. I did not fear we should out-run the Constable; for though we should Pamper our selves for a whole Week; we could not spend a *Sonnet*, and I have Four about me, besides Two *Epigrams*, Two *Odes*, and an *Eclogue*. Would to God, said I, it were so in our World; for I know a good many honest Poets there, who are ready to Starve, and who might live plentifully, if that Money would pass in Payment. I farther asked him, If these Verses would always serve, if one Transcribed them? He made answer, No, and so went on. When an Author has Composed any; he carries them to the Mint, where the sworn Poets of the Kingdom sit in Court. There these versifying Officers essay the pieces; and if they be judged Sterling, they are rated not according to their Coyn; that's to say, That a *Sonnet* is not always as good as a *Sonnet*; but according to the intrinsick value of the piece; so that if any one Starve, he must be a Blockhead: For Men of Wit make always good Chear. With Extasie, I was admiring the judicious Policy of that Country, when he proceeded in this manner: There are others who keep Publick-house, after a far different manner: When one is about to be gone, they demand proportionably to the Charges, an Acquittance for the other World; and when that is given them, they write down in a great Register, which they call *Doomsday's Book*, much after this manner. *Item*, The value of so many Verses, delivered such a Day, to such a Person, which he is to pay upon the receipt of this Acquittance, out of his readiest Cash: And when they find themselves in danger of Death, they cause these Registers to be Chopt in pieces, and swallow them down; because they believe,

that

at if they were not thus digested, they ould be good for nothing.

This Conversation was no hinderance to ur Journey; for my Four-legged Porter gged on under me, and I rid ſtradling n his Back. I ſhall not be particular in re- ting to you, all the Adventures, that appened to us on our way, till we arri- ed at length at the Town, where the ing holds his Reſidence. I was no ooner come, but they carryed me to the alace, where the Grandees received me ith more Moderation, than the People ad done, as I paſſed the Streets: But oth great and ſmall concluded, That, ithout doubt, I was the Female of the ueen's little Animal. My Guide was y Interpreter; and yet he himſelf un- erſtood not the Riddle, and knew not hat to make of that little Animal of ie Queen's; but we were ſoon ſatis- ed as to that; for the King having ſome me conſidered me, ordered it to be rought, and about half an hour after, ſaw a company of Apes, wearing Ruffs nd Breeches, come in, and amongſt them a ttle Man, almoſt of my own Built, for he ent on Two Legs; ſo ſoon as he per- eived me, he Accoſted me with a *Cria- o de vueſtra merced*. I anſwered his reeting, much in the ſame Terms. But

alas!

alas! no sooner had they seen us talk together, but they believed their Conjecture to be true; and so, indeed, it seemed; for he of all the By-standers, that past the most favourable Judgment upon us, protested, that our Conversation, was a Chattering we kept for Joy at our meeting again. That little Man told me, that he was an *European*, a Native of old *Castille*: That he had found a means by the help of Birds, to mount up to the World of the Moon, where then we were: That falling into the Queen's Hands, she had taken him for a Monkey, because Fate would have it so: That in that Country they cloath Apes in a *Spanish* Dress; and that upon his arrival, being found in that habit, she had made no doubt, but he was of the same kind. It could not otherwise be, replied I, but having tried all Fashions of Apparel upon them, none were found so Ridiculous, and by consequence more becoming a kind of Animals, which are only entertained for Pleasure and Diversion. That shews you little understand the Dignity of our Nation, answered he, for whom the Universe breeds Men, only to be our Slaves, and Nature produces nothing but objects of Mirth and Laughter. He then intreated me to tell him, how I durst be so bold, as to Scale the Moon
with

ith the Machine I told him of? I an-
ered, That it was because he had carried
vay the Birds, which I intended to have
ade use of. He smiled at this Raillery;
d about a quarter of an hour after, the
ing commanded the Keeper of the Mon-
ys to carry us back, with express Orders
make the *Spaniard* and me lie together,
at we might procreate a breed of Apes
his Kingdom. The King's Pleasure
as punctually obeyed; at which I was
ry glad, for the satisfaction I had, of hang a Mate to converse with, during the
litude of my Brutification. One Day
y Male (for I was taken for the Female)
ld me, That the true reason, which had
liged him to travel all over the Earth,
d at length to abandon it for the Moon,
as, that he could not find so much as
e Country, where even Imagination
as at liberty. Look ye, said he, how
e Wittiest thing you can say, unless you
ear a Cornered Cap, if it thwart the
inciples of the Doctors of the Robe,
u are an Ideot, a Fool, and something
orse, perhaps. I was about to have been
it into the Inquisition at home, for
aintaining to the Pedants Teeth, That
ere was a *Vacuum*, and that I knew no
e matter in the World, more Ponderous
an another. I asked him, what probable

E 4 Argu-

Arguments he had, to confirm so new an Opinion? To evince that, answered he, you must suppose that there is but one Element; for though we see Water, Earth, Air and Fire distinct, yet are they never found to be so perfectly pure, but that there still remains some Mixture. For example, When you behold Fire, it is not Fire but Air much extended; the Air is but Water much dilated; Water is but liquified Earth, and the Earth it self, but condensed Water; and thus if you weigh Matter seriously, you'll find it is but one, which like an excellent Comedian here below acts all Parts, in all sorts of Dresses: Otherwise we must admit as many Elements, as there are kinds of Bodies: And if you ask me why Fire burns, and Water cools, since it is but one and the same matter, I answer, That that matter acts by Sympathy, according to the Disposition it is in, at the time when it acts. Fire which is nothing but Earth also, more dilated than is fit for the constitution of Air, strives to change into it self, by Sympathy, whatever it meets with: Thus the heat of Coals, being the most subtile Fire, and most proper to penetrate a Body, at first slides through the pores of our Skin; and because it is a new matter that fills us, it makes us exhale in Sweat;
that

that Sweat dilated by the Fire is converted to a Steam, and becomes Air; that Air being farther rarified by the heat of the *Antiperistasis*, or of the Neighbouring Stars, is called Fire, and the Earth abandoned by the Cold and Humidity, which were Ligaments to the whole, falls to the ground: Water, on the other hand, though it no ways differ from the matter of Fire, but in that it is closer, burns us not; because that being dense by Sympathy, it closes up the Bodies it meets with, and the Cold we feel is no more, but the effect of our Flesh contracting it self, because of the Vicinity of Earth or Water, which constrains it to a Resemblance. Hence it is, that those who are troubled with a Dropsie, convert all their nourishment into Water; and the Cholerick convert all the Blood, that is formed in their Liver, into Choler. It being then supposed, that there is but one Element; it is most certain, that all Bodies, according to their several qualities, incline equally towards the Center of the Earth.

But you'll ask me, Why then does Iron, Metal, Earth and Wood, descend more swiftly to the Center than a Sponge, if it be not that it is full of Air, which naturally tends upwards? That is not at all the Reason, and thus I
make

make it out: Though a Rock fall with greater Rapidity than a Feather, both of them have the same inclination for the Journey; but a Cannon Bullet, for instance, where the Earth pierced through, would precipitate with greater haste to the Center thereof, than a Bladder full of Wind; and the reason is, because that mass of Metal, is a great deal of Earth contracted into a little space, and that Wind a very little Earth in a large space: For all the parts of Matter, being so closely joined together in the Iron, encrease their force by their Union; because being thus compacted, they are many that Fight against a few, seeing a parcel of Air equal to the Bullet in Bigness, is not equal in Quantity.

Not to insist on a long Deduction of Arguments to prove this, tell me in good earnest, How a Pike, a Sword or a Dagger wound us? If it be not, because the Steel, being a matter, wherein the parts are more continuous, and more closely knit together, than your Flesh is, whose Pores and Softness shew, that it contains but very little Matter, within a great extent of Place; and that the point of the Steel that pricks us, being almost an innumerable number of Particles of matter, against a very little Flesh, it forces it to yeild to the stronger,

in

in the same manner as a Squadron in close order, will easily break through a more open Battallion; for why does a Bit of red hot Iron, burn more than a Log of Wood all on Fire? Unless it be, that in the Iron, there is more Fire in a small space, seeing it adheres, to all the parts of the Metal, than in the Wood which being very Spongy, by consequence contains a great deal of *Vacuity*; and that *Vacuity*, being but a Privation of Being, cannot receive the form of Fire. But, you'll object, you suppose a *Vacuum*, as if you had proved it, and that's begging of the question: Well then I'll prove it, and though that difficulty be the Sister of the *Gordian knot*, yet my Arms are strong enough to become its *Alexander*.

Let that vulgar Beast, then, who does not think it self a Man, had it not been told so, answer me if it can: Suppose now there be but one Matter, as I think I have sufficiently peoved; whence comes it, that according to its Appetite, it enlarges or contracts its self; whence is it, that a piece of Earth, by being Condensed becomes a Stone? Is it that the parts of that Stone are placed one with another, in such a manner, that wherever that grain of Sand is settled, even there, or in the same point, another grain of Sand is Lodged? That

cannot

cannot be, no not according to their own Principles, seeing there is no Penetration of Bodies: But that matter must have crowded together, and if you will, abridged it self, so that it hath filled some place which was empty before. To say, that it is incomprehensible, that there should be a Nothing in the World, that we are in part made up of Nothing: Why not, pray? Is not the whold World wrapt up in Nothing? Since you yield me this point, then confess ingeniously, that it's as rational, that the World should have a Nothing within it, as Nothing about it.

I well perceive you'll put the question to me, Why Water compressed in a Vessel by the Frost should break it, if it be not to hinder a Vacuity? But I answer, That that only happens, because the Air overhead, which as well as Earth and Water, tends to the Center, meeting with an empty Tun by the way, takes up his Lodging there: If it find the pores of that Vessel, that's to say, the ways that lead to that void place, too narrow, too long, and too crooked, with impatience it breaks through and arrives at its Tun.

But not to trifle away time, in answering all their objections, I dare be bold to say, That if there were no *Vacuity*, there could be no Motion; or else a Penetration of Bodies

dies muſt be admitted; for it would be a little too ridiculous to think, that when a Gnat puſhes back a parcel of Air with its Wings, that parcel drives another before it, that other another ſtill; and that ſo the ſtirring of the little Toe of a Flea, ſhould raiſe a bunch upon the Back of the Univerſe. When they are at a ſtand, they have recourſe to Rarefaction: But in good earneſt, How can it be when a Body is rarified, that one Particle of the Maſs does recede from another Particle, without leaving an empty Space betwixt them; muſt not the two Bodies, which are juſt ſeparated, have been at the ſame time in the ſame place of this; and that ſo they muſt have all three penetrated each other? I expect you'll ask me, why through a Reed, a Syringe or a Pump, Water is forced to aſcend contrary to its inclination? To which I anſwer, That that's by violence, and that it is not the fear of a *Vacuity*, that turns it out of the right way; but that being linked to the Air by an imperceptible Chain, it riſes when the Air, to which it is joined, is raiſed.

That's no ſuch knotty Difficulty, when one knows the perfect Circle, and the delicate Concatenation of the Elements: For if you attentively conſider the Slime, which joines the Earth and Water together in Marriage,

riage, you'll find that it is neither Earth nor Water; but the Mediator betwixt these Two Enemies. In the same manner, the Water and Air reciprocally send a Mist, that dives into the Humours of both, to negotiate a Peace betwixt them; and the Air is reconciled to the Fire, by means of an interposing Exhalation which Unites them.

I believe he would have proceeded in his Discourse, had they not brought us our Victuals; and seeing we were a hungry, I stopt my Ears to his discourse, and opened my Stomack to the Food they gave us.

I remember another time, when we were upon our Philosophy, for neither of us took pleasure to Discourse of mean things: I am vexed, said he, to see a Wit of your stamp, infected with the Errors of the Vulgar. You must know then, in spight of the Pedantry of *Aristotle*, with which your Schools in *France* still ring, That every thing is in every thing; that's to say, for instance, That in the Water there is Fire, in the Fire Water, in the Air Earth, and in the Earth Air: Though that Opinion makes Scholars open their Eyes as big as Sawcers, yet it is easier to prove it, than perswade it. For I ask them, in the first place, if Water does not breed Fish: If they deny it, let them dig a Pit, fill it with meer Element,

ment, and to prevent all blind Objections, let them if they please, strain it through a Strainer, and I'll oblige my self, in case they find no Fish therein, within a certain time, to drink up all the Water they have poured into it: But if they find Fish, as I make no doubt on't; it is a convincing Argument, that there is both Salt and Fire there. Consequentially now, to find Water in Fire; I take it to be no difficult Task. For let them chuse Fire, even that which is most abstracted from Matter, as Comets are, there is a great deal in them still; seeing if that Unctuous Humour, whereof they are engendred, being reduced to a Sulphur, by the heat of the Antiperistasis which kindles them, did not find a curb of its Violence in the humid Cold, that qualifies and resists it, it would spend it self in a trice like Lightning. Now that there is Air in the Earth, they will not deny it; or otherwise they have never heard of the terrible Earth-quakes, that have so often shaken the Mountains of *Sicily*: Besides, the Earth is full of Pores, even to the least grains of Sand that compass it. Nevertheless, no Man hath as yet said, that these Hollows were filled with *Vacuity*: It will not be taken amiss then, I hope, if the Air takes up its quarters there. It remains to be proved, that there is Earth in the

the Air; but I think it scarcely worth my pains, seeing you are convinced of it, as often as you see, such numberless Legions of Atomes fall upon your heads, as even stiffle Arithmetick.

But let us pass from simple to compound Bodies, they'll furnish me with much more frequent Subjects; and to demonstrate that all things are in all things, not that they change into one another, as your *Peripateticks* Juggle, (for I will maintain to their Teeth, that the Principles mingle, separate, and mingle again in such a manner, that that hath been made Water by the Wise Creator of the World, will always be Water:) I shall suppose no Maxime, as they do, but what I prove.

And therefore take a Billet, or any other combustible stuff, and set Fire to it, they'll say when it is in a Flame, That what was Wood is now become Fire; but I maintain the contrary, and that there is no more Fire in it, when it is all in Flame, than before it was kindled; but that which before was hid in the Billet, and by the Humidity and Cold hindered from acting; being now assisted by the Stronger, hath rallied its forces against the Phlegm that choaked it, and commanding the Field of Battle, that was possessed by its Enemy, triumphs over his Jaylor, and appears without Fetters.

ers. Don't you see how the Water flees out at the two ends of the Billet, hot and smoaking from the Fight it was engaged in. That flame which you see rise on high, is the purer Fire, unpestered from the Water, and by consequence the readiest to return home to it self: Nevertheless it Unites it self, by tapering into a Piramide, till it rise to a certain height, that it may pierce through the thick Humidity of the Air, which resists it; but as in mounting it disengages it self by little and little, from the violent company of its Landlords; so it diffuses it self, because then it meets with nothing that thwarts its passage, which negligence, though, is many times the cause of a second Captivity: For marching stragglingly, it wanders sometimes into a Cloud, and if it meet there with a Party of its own, sufficient to make head against a Vapour; they Engage, Grumble, Thunder and Roar, and the Death of Innocents is many times the effect of the animated Rage, of those inanimate Things. If when it finds it self pestered, among those crudities of the middle Region, it is not strong enough to make a defence; it yields to its Enemy upon discretion, which by its weight, constrains it to fall again to the Earth: And this Wretch, inclosed in a drop of Rain, may, perhaps, fall at the Foot of an Oak,

Oak, whose Animal Fire will invite the poor Straggler, to take a Lodging with him; and thus you have it in the same condition again, as it was a few Days before.

But let us trace the Fortune of the other Elements, that composed that Billet. The Air retreats to its own Quarters also, though blended with Vapours; because the Fire all in a rage, drove them briskly out *Pell-mell* together. Now you have it serving the Winds for a Tennis-ball, furnishing Breath to Animals, filling up the Vacuities that Nature hath left; and, it may be, also wrapt up in a drop of Dew, suckling the thirsty Leaves of that Tree, whither our Fire retreated: The Water driven from its Throne by the Flame, being by the heat elevated to the Nursery of the Meteors, will distil again in Rain upon our Oak, as soon as upon another; and the Earth being turned to Ashes, and then cured of its Sterility, either by the nourishing Heat of a Dunghill, on which it hath been thrown, or by the vegetative Salt of some neighbouring Plants, or by the teeming Waters of some Rivers, may happen also to be near this Oak, which by the heat of its Germ, will attract it, and convert it into a part of its bulk.

World of the Moon.

In this manner, these Four Elements undergo the same Destiny, and return to the same State, which they quitted but a few days before: So that it may be said, that all that's necessary for the composition of a Tree, is in a Man; and in a Tree, all that's necessary for making of a Man. In fine, according to this way, all things will be found in all things; but we want a *Prometheus*, to pluck us out of the Bosom of Nature, and render us sensible, which I am willing to call the *First Matter*.

These were the things, I think, with which we past the time; for that little *Spaniard* had a quaint Wit. Our conversation, however, was only in the Night time; because from Six a clock in the morning until night, Crowds of the People that came to stare at us in our Lodging, would have disturbed us: For some threw us Stones, others Nuts, and others Grass; there was no talk, but of the Kings Beasts; we had our Victuals daily at set hours, and the King and Queen took the pains often to feel my Belly, to see if I did not begin to swell, for they had an extraordinary desire to have a Race of these little Animals. I cannot tell, whether it was that I minded their Gestures and Tones, more than my Male did: But I learnt sooner than he, to understand their Language, and to smatter a little in it, which made us to be lookt upon, in another guefs manner than formerly; and the news thereupon flew prefently all over the Kingdom, that two Wild Men had been found, who were lefs than other Men, by reafon of the bad Food we had had in the Defarts; and who through a defect of their Parents Seed, had not the fore Legs ftrong enough to fupport their Bodies.

This belief would have taken rooting, by being fpread, had it not been for the Learned Men of the Country, who oppofed it, faying, That it was horrid Impiety to believe, not only Beafts, but Monfters to be of their kind. It would be far more probable, (added the calmer Sort) that our Domeftick Beafts, fhould participate of the priviledge of Humanity, and by confequence of Immortality, as being bred in our Country, than a Monftrous Beaft, that talks of being born I know not where, in the Moon; and then obferve the difference betwixt us and them. We walk upon Four Feet, becaufe God would not truft fo precious a thing, upon weaker Supporters, and he was afraid leaft marching otherwife, fome Mifchance might befall Man; and therefore he took the pains, to reft him upon four Pillars, that he might not fall, but difdaining to have a hand in the

the Fabrick of these two Brutes, he left them to the Caprice of Nature, who not concerning her self with the loss of so small a matter, supported them only by Two Feet.

Birds themselves, said they, have not had so hard measure as they; for they have got Feathers, at least to supply the weakness of their Legs, and to cast themselves in the Air, when we pursue them; whereas Nature, depriving these Monsters of Two Legs, hath disabled them from scaping our Justice.

Besides, consider a little how they have the Head raised toward Heaven, it is because God would punish them with scarcity of all things, that he hath so placed them; for that supplicant Posture, shews that they complain to Heaven of him that Created them, and that they beg Permission to make their best of our Leavings. But we, on the contrary, have the Head bending downwards, to behold the Blessings, whereof we are the Masters, and as if there were nothing in Heaven, that our happy condition needed Envy.

I heard such Discourses, or the like, daily at my Lodge; and at length they so curbed the minds of the people, as to that point, that it was decreed, That at best, I should only pass for a Parrot without Feathers;

for they confirmed thofe, who were already perfwaded, in that I had but two Legs no more than a Bird, which was the caufe that I was put into a Cage, by expreſs orders from the Privy Council.

There the Queen's Bird-keeper, taking the pains daily to teach me to Whiſtle, as they do Stares or Singing-Birds here, I was really happy in that I wanted not Food: In the mean while, with the Sonnets the Spectators ſtunned me, I learnt to ſpeak as they did; ſo that when I was got to be ſo much Maſter of the Idiom, as to expreſs moſt of my thoughts, I told them the fineſt of my Conceits. The Quaintneſs of my Sayings, was already the entertainment of all Societies, and my Wit was ſo much eſteemed, that the Council was obliged to Publiſh an Edict, forbidding all People to believe, that I was endowed with Reaſon; with expreſs Commands to all Perſons, of what Quality or Condition ſoever, not to imagine, but that whatever I did, though never ſo wittily, proceeded only from Inſtinct.

Nevertheleſs, the deciſion of what I was, divided the Town into Two Factions. The party that ſtood for me, encreaſed daily; and at length in ſpight of the *Anathema*, whereby they endeavoured to ſcare the multitude: They who held for me, demanded ed a Convention of the States, for determining that Controverſie. It was long before they could agree, in the Choice of thoſe who ſhould have a Vote; but the Arbitrators pacified the heat, by making the number of both parties equal, who ordered that I ſhould be brought unto the Aſſembly, as I was: But I was treated there with all imaginable Severity. My Examiners, amongſt other things, put queſtions of Philoſophy to me; I ingenuouſly told them, all that my Tutor had heretofore taught me, but they eaſily refuted me by more convincing Arguments: So that having nothing to anſwer for my ſelf, my laſt refuge was to the Principles of *Ariſtotle*, which ſtood me in as little ſtead, as his Sophiſms did; for in two Words, they let me ſee the falſity of them. That ſame *Ariſtotle*, ſaid they, whoſe Learning you brag ſo much of, did without doubt accommodate Principles to his Philoſophy; inſtead of accommodating his Philoſophy to Principles; and beſides he ought to have proved them, at leaſt to be more rational, than thoſe of the other Sects, you mentioned to us: Wherefore the good Man will not take it ill, we hope, if we bid him God b'w'. In fine, when they perceived that I did nothing but bawl, that they were not more knowing than *Ariſtotle*, and that I was forbid to difpute againſt thoſe who denied his Principles: They all unanimouſly concluded, That I was not a Man, but, perhaps, a kind of *Eſtriage*, ſeeing I carried my Head upright like them, that I walked on two Legs; and that, in ſhort, but for a little Down, I was every way like one of them, ſo that the Bird-keeper was ordered to have me back to my Cage. I ſpent my time pretty pleaſantly there, for becauſe I had correctly learned their Language, the whole Court took pleaſure to make me prattle. The Queen's Maids, among the reſt, ſlipt always ſome Boon into my Basket; and the gentileſt of them all, having conceived ſome kindneſs for me, was ſo tranſported with Joy, when in private I entertained her, with the manners and divertiſements of the People of our World, and eſpecially our Bells, and other Inſtruments of Muſick, that ſhe proteſted to me, with Tears in her Eyes, That if ever I found my ſelf in a condition to fly back again to our World, ſhe would follow me with all her Heart.

One Morning early, having ſtarted out of my Sleep, I found her Taboring upon the grates of my Cage: Take good heart, ſaid ſhe to me, yeſterday in Council, a War was reſolved upon, againſt the King ───── I hope that during the hurry

hurry of Preparations, whilſt our Monarch and his Subjects are abſent, I may find an occaſion to make your eſcape. How, a War, ſaid I interruping her, have the Princes of this World, then, any quarrels amongſt themſelves, as thoſe of ours have? Good now, let me know their way of Fighting.

When the Arbitrators, replied ſhe, who are freely choſen by the two Parties, have appointed the time for raiſing Forces for their March, the number of Combatants, the day and place of Battle, and all with ſo great equality, that there is not one Man more in one Army, than in the other: All the maimed Soldiers on the one ſide, are liſted in one Company; and when they come to engage, the *Mareſhalls de Camp*, take care to expoſe them to the maimed of the other ſide: The Giants are matched with Coloſſes, the Fencers with thoſe that can handle their Weapons, the Valiant with the Stout, the Weak with the Infirm, the Sick with the Indiſpoſed, the Sturdy with the Strong; and if any undertake to ſtrike at another, than the Enemy he is matched with, unleſs he can make it out, that it was by miſtake, he is Condemned for a Coward. When the Battle is over, they take an account of the Wounded, the Dead and the Priſoners,

ners, for Run-aways they have none; and if the loſs be equal on both ſides, they draw Cuts, who ſhall be Proclaimed Victorious.

But though a Kingdom hath defeated the Enemy in open War, yet there is hardly any thing got by it; for there are other ſmaller Armies of Learned and Witty Men, on whoſe Diſputations the Triumph or Servitude of States wholly depends.

One Learned Man grapples with another, one Wit with another, and one Judicious Man with another Judicious Man: Now the Triumph which a State gains in this manner, is reckoned as good as three Victories by open force. After the Proclamation of Victory, the Aſſembly is broken up, and the Victorious People, either chuſe the Enemies King to be theirs, or confirm their own.

I could not forbear to Laugh, at this ſcrupulous way of giving Battle; and for an Example of much ſtronger Politicks, I alledged the Cuſtoms of our *Europe*, where the Monarch would be ſure, not to let ſlip any favourable occaſion of gaining the day; but mind what ſhe ſaid as to that.

Tell me, pray, if your Princes uſe not a pretext of Right, when they levy Arms: No doubt, anſwered I, and of the

the Juſtice of their Cauſe too. Why then, replied ſhe, do they not chuſe Impartial and Unſuſpected Arbitrators to compoſe their Differences? And if it be found, that the one has as much Right as the other, let things continue as they were; or let them play a game at *Picket*, for the Town or Province that's in diſpute.

But why all theſe Circumſtances, replied I, in your way of Fighting? Is it not enough, that both Armies are equal in the number of Men? Your Judgment is Weak, anſwered ſhe. Would you think in Conſcience, that if you had the better of your Enemy, Hand to Hand, in an open Field, you had fairly overcome him, if you had had on a Coat of Mail, and he none; if he had had but a Dagger, and you a Tuck; and in a Word, if he had had but one Arm, and you both yours? Nevertheleſs, what Equality ſoever you may recommend to your Gladiators, they never fight on even terms; for the one will be a tall Man, and the other Short; the one skilful at his Weapon, and the other a Man that never handled a Sword; the one will be ſtrong, and the other Weak: And though theſe Diſproportions were not, but that the one were as skillful and ſtrong as the other; yet ſtill they might not be rightly
matched;

matched; for one, perhaps, may have more Courage than the other, who being rash and hot-headed, inconcerned in danger, as not foreseeing it, of a bilious Temper, a more contracted Heart, with all the qualities that constitute Courage, (as if that, as well as a Sword, were not a Weapon which his Adversary hath not:) He makes nothing of falling desperately, upon terrifying and killing this poor Man, who foresees the danger; has his Heat choked in Phlegme, and a Heart too wide to close in the Spirits in such a posture, as is necessary for thawing that Ice, which is called Cowardise. And now you praise that Man, for having killed his Enemy at odds, and praising him for his Boldness, you praise him for a Sin against nature; seeing such Boldness tends to its destruction. And this puts me in mind to tell ye, that some Years ago, application was made to the Council of War, for a more circumspect and conscientious Rule to be made, as to the way of Fighting. The Philosopher who gave the advice, if I mistake it not, spake in this manner.

You imagine, Gentlemen, that you have very equally balanced the advantages of two Enemies, when you have chosen both Tall Men, both skillful, and both couragious: But that's not enough, seeing after all

all, the Conquerour muſt have the better on't, either through his Skill, Strength, or good Fortune. If it be by Skill, without doubt he hath taken his Adverſary on the blind ſide, which he did not expect; or ſtruck him ſooner than was likely, or faining to make his Paſs on one ſide, he hath attacked him on the other: Nevertheleſs all this is Cunning, Cheating, and Treachery, and none of theſe make a brave Man: If he hath triumphed by Force, would you judge his Enemy over-come, becauſe he hath been over-powered? No; doubtleſs, no more than you'll ſay, that a Man hath loſt the Victory, when overwhelm'd by a Mountain, it was not in his power to gain it: Even ſo, the other was not overcome, becauſe he was not in a ſuitable Diſpoſition, at that nick of time, to reſiſt the violences of his Adverſary. If Chance hath given him the better of his Enemy, Fortune ought then to be Crowned, ſince he hath contributed nothing to it; and, in fine, the vanquiſhed is no more to be blamed, than he who at Dice having thrown Seventeen, is beat by another that throws three Sixes.

They confeſſed he was in the right; but that it was impoſſible, according to humane Appearances, to remedy it; and that it was better, to ſubmit to a ſmall inconvenience, than

than to open a door to a hundred of greater Importance.

She entertained me no longer at that time, becauſe ſhe was afraid to be found alone with me ſo early; not that Impudicity is a Crime in that Country: On the contrary, except Malefactors Convicted, all Men have power over all Women; and in the ſame manner, a Woman may bring her Action againſt a Man for refuſing her: But ſhe durſt not keep me company publickly, becauſe the Members of Council, at their laſt meeting, had ſaid, That it was chiefly the Women who gave it out that I was a Man, that under that pretext they might cover the violent deſire they had of enjoying Beaſts, and of committing with me ſins againſt Nature, without a bluſh; which was the reaſon, that for a long time, I neither ſaw her, nor any other of her Sex.

In the mean time, ſome muſt needs have revived the Diſputes, about the Definition of my Being; for whilſt I was thinking of nothing elſe, but of dying in my Cage, I was once more brought out to have another Audience. I was then queſtioned, in preſence of a great many Courtiers, upon ſome points of Natural Philoſophy; and, as I take it, my Anſwers gave ſome kind of Satisfaction;

on; for the President declared to me at large, his thoughts concerning the structure of the World. They seemed to me very ingenious; and had he not traced it to its Original, which he maintained to be Eternal, I should have thought his Philosophy more rational than our own: But as soon as I heard him maintain a Foppery, so contrary to our Faith, I broke with him; at which he did but laugh; and that obliged me to tell him, That since they were thereabouts with it, I began again to think, that their World was but a Moon. But then all cried, Don't you see here Earth, Rivers, Seas? what's all that then? No matter, said I, *Aristotle* assures us it is but a Moon; and if you had said the contrary in the Schools, where I have been bred, you would have been hissed at. At this they all burst out in laughter; you need not ask, if it was their Ignorance, that made them do so; for in the mean time I was carried back to my Cage.

But some more passionate Doctors, being informed, that I had the boldness to affirm, That the Moon, from whence I came, was a World; and that their World was no more but a Moon, thought it might give them a very just pretext, to have me condemned to the Water; for that's that's their way of rooting out Hereticks. For that end, they went in a Body, and complained to the King, who promised them Justice; and order'd me once more to be brought to the Bar.

Now was I the third time Un-caged; and then the most Ancient spoke, and pleaded against me. I do not well remember his Speech; because I was too much frighted, to receive the tones of his Voice without disorder; and because also in declaiming, he made use of an Instrument, which stunn'd me with its noise: It was a Speaking-Trumpet, which he had chosen on purpose, that by its Martial Sound he might rouse them to my death; and by that Emotion of their Spirits, hinder Reason from performing its Office: As it happens in our Armies, where the noise of Drums and Trumpets, hinders the Souldiers from minding the importance of their Lives. When he had done, I rose up to defend my Cause; but I was excused from it, by an Accident that will surprize you. Just as I had opened my Mouth, a Man, who with much ado had pressed through the Crowd, fell at the King's Feet, and a long while rouled himself upon his Back in his presence. This practice did not at all surprize me, because I knew it to be the posture they put them-

themselves into, when they have a mind to be heard in publick: I only stopt my own Harangue, and gave Ear to his.

Just Judges, said he, listen to me; you cannot Condemn that Man, that Monkey or Parrot, for saying, That the Moon, from whence he comes, is a World; for if he be a Man, though he were not come from the Moon, since all Men are free, is not he free also, to imagine what he pleases? How, can you constrain him not to have Visions, as well as you? You may very well force him to say, That the Moon is not a World, but he will not believe it for all that; for to believe a thing, some possibilities enclining more to the Yea, than to the Nay, must offer to ones Imagination: And unless you furnish him with that Probability, or his own mind hit upon it, he may very well tell you, that he believes, but still remain an Infidel.

I am now to prove, that he ought not to be condemned, if you list him in the Catalogue of Beasts.

For suppose him to be an Animal without Reason, would it be rational in you to Condemn him, for offending against it? Ie hath said, that the Moon is a World. Now Beasts act only by the instinct of Nature: it is Nature then that says so, and not he: To think that wise Nature, who

G   hath

hath made the World, and the Moon knows not her self what it is; and that ye who have no more Knowledge, but what ye derive from her, should more certainly know it, would be very Ridiculous. But if Passion should make you renounce your Principles, and you should suppose, that Nature does not guide Beasts; blush, at least, to think on't, that the Caprices of a Beast should so discompose you. Really, Gentlemen, should you meet with a Man come to the Years of Discretion, who made it his business to inspect the Government of *Pismires*, giving a blow to one that had overthrown its Companion, imprisoning another that had robb'd its Neighbour of a grain of Corn, and inditing a third, for leaving its Eggs; would you not think him a mad Man, to be employed in things so far below him, and to pretend to give Laws to Animals, that never had Reason? How will you then, most Venerable Assembly, justifie your selves, for being so concerned at the Caprices of that little Animal? Just Judges, I have no more to say.

When he had made an end, all the Hall rung again with a kind of Musical Applause; and after all the Opinions had been canvased, during the space of a large quarter of an hour, the King gave Sentence:

That

That for the future, I should be reputed to be a Man, accordingly set at liberty; and that the Punishment of being Drowned, should be converted into a publick Disgrace (the most honourable way of satisfying the Law in that Country) whereby I should be obliged to retract openly, what I had maintained, in saying, That the Moon was a World, because of the Scandal, that the novelty of that opinion, might give to weak Brethren.

This Sentence being pronounced, I was taken away out of the Palace, richly Cloathed; but in derision, carried in a magnificent Chariot, as on a Tribunal; which four Princes in Harness drew, and in all the publick places of the Town, I was forced to make this Declaration:

Good People, I declare to you, That this Moon here is not a Moon, but a World; and that that World below is not a World, but a Moon: This the Council thinks fit you should believe. After I had Proclaimed this, in the five great places of the Town; my Advocate came and reached me his Hand to help me down. I was in great amaze, when after I had Eyed him, I found him to be my Spirit; we were an hour in embracing one another: Come lodge with me, said he, for if you return to Court, after a Publick Disgrace,

you will not be well lookt upon : Nay more, I muſt tell you, that you would have been ſtill amongſt the Apes yonder, as well as the *Spaniard* your Companion ; if I had not in all Companies publiſhed the vigour and force of your Wit, and gained from your Enemies, the protection of the great Men, in your favours. I ceaſed not to thank him all the way, till we came to his Lodgings ; there he entertained me till Supper-time, with all the Engines he had ſet a work to prevail with my Enemies. Notwithſtanding, the moſt ſpecious pretexts they had uſed for riding the Mobile, to deſiſt from ſo unjuſt a Proſecution : But as they came to acquaint us, that Supper was upon the Table, he told me, that to bear me company that evening, he had invited Two Profeſſors of the Univerſity of the Town to Sup with him : I'll make them, ſaid he, fall upon the Philoſophy, which they teach in this World, and by that means you ſhall ſee my Landlord's Son : He's as Witty a Youth as ever I met with ; he would prove another *Socrates*, if he could uſe his Parts aright, and not bury in Vice the Graces wherewith God continually viſits him, by affecting a Libertiniſm, as he does, out of a Chimerical Oſtentation and Affectation of the name of a Wit. I have taken Lodgings

ings here, that I may lay hold on all Opportunities of Inftructing him: He faid no more, that he might give me the Liberty to fpeak, if I had a mind to it; and then made a fign, that they fhould ftrip me of my difgraceful Ornaments, in which I ftill gliftered.

The Two Profeffors, whom we expected, entered, juft as I was undreft, and we went to fit down to Table, where the Cloth was laid, and where we found the Youth, he had mentioned to me, fallen to already. They made him a low Reverence, and treated him with as much refpect, as a Slave does his Lord. I asked my Spirit the reafon of that, who made me anfwer, that it was becaufe of his Age; feeing in that World, the Aged rendered all kind of Refpect and Difference to the Young; and which is far more, that the Parents obeyed their Children, fo foon as by the Judgment of the Senate of Philofophers, they had attained to the Years of Difcretion. You are amazed, continued he, at a Cuftom fo contrary to that of your Country; but it is not all repugnant to Reafon: For fay, in your Confcience, when a brisk young Man is at his Prime in Imagining, Judging, and Acting, is not he fitter to govern a Family, than a Decrepit piece of

Threescore Years, dull and doting, whose Imagination is frozen under the Snow of Sixty Winters, who follows no other Guide, but what you call, the Experience of happy Successes; and yet are no more, but the bare effects of Chance, against all the Rules and Oeconomy of humane Prudence? And as for Judgment, he hath but little of that neither, though the people of your World make it the Portion of Old Age: But to undeceive them, they must know, That that which is called Prudence in an Old Man, is no more but a panick Apprehension, and a mad Fear of acting any thing, where there is danger: So that when he does not run a Risk, wherein a Young Man hath lost himself; it is not that he foresaw the Catastrophe, but because he had not Fire enough to kindle those noble Flashes, which make us dare: Whereas the Boldness of that Young Man, was as a pledge of the good Success of his design; because the same Ardour that speeds and facilitates the execution, thrust him upon the undertaking. As for Execution, I should wrong your Judgment; if I endeavoured to convince it by proofs: You know that Youth alone is proper for Action; and were you not fully perswaded of this, tell me, pray,

when

when you respect a Man of Courage, is it not because he can revenge you on your Enemies or Oppressors? And does any thing, but meer Habit, make you consider him, when a Battalion of Seventy *Januarys* hath frozen his Blood, and chilled all the noble Heats that youth is warmed with? When you yeild to the Stronger, is it not that he should be obliged to you for a Victory, which you can Dispute him? Why then should you submit to him, when Laziness hath softened his Muscles, weakened his Arteries, evaporated his Spirits, and suckt the Marrow out of his Bones? If you adore a Woman, is it not because of her Beauty? Why should you then continue your Cringes, when Old Age hath made her a Ghost, which only represents a hideous Picture of Death? In short, When you loved a Witty Man, it was because by the Quickness of his Apprehension, he unravelled an intricate Affair, seasoned the choicest Companies with his quaint Sayings, and sounded the depth of Sciences with a single Thought; and do you still honour him, when his worn Organs disappoint his weak Noddle, when he is become dull and uneasy in Company, and when he looks like an aged Fairy rather than a rational Man?

Threescore Years, dull and doting, whose Imagination is frozen under the Snow of Sixty Winters, who follows no other Guide, but what you call, the Experience of happy Successes; and yet are no more, but the bare effects of Chance, against all the Rules and Oeconomy of humane Prudence? And as for Judgment, he hath but little of that neither, though the people of your World make it the Portion of Old Age: But to undeceive them, they must know, That that which is called Prudence in an Old Man, is no more but a panick Apprehension, and a mad Fear of acting any thing, where there is danger: So that when he does not run a Risk, wherein a Young Man hath lost himself; it is not that he foresaw the Catastrophe, but because he had not Fire enough to kindle those noble Flashes, which make us dare: Whereas the Boldness of that Young Man, was as a pledge of the good Success of his design; because the same Ardour that speeds and facilitates the execution, thrust him upon the undertaking. As for Execution, I should wrong your Judgment; if I endeavoured to convince it by proofs: You know that Youth alone is proper for Action; and were you ont fully perswaded of this, tell me, pray,

when

when you respect a Man of Courage, is it not because he can revenge you on your Enemies or Oppressors? And does any thing, but meer Habit, make you consider him, when a Battalion of Seventy *Januarys* hath frozen his Blood, and chilled all the noble Heats that youth is warmed with? When you yeild to the Stronger, is it not that he should be obliged to you for a Victory, which you can Dispute him? Why then should you submit to him, when Laziness hath softened his Muscles, weakened his Arteries, evaporated his Spirits, and suckt the Marrow out of his Bones? If you adore a Woman, is it not because of her Beauty? Why should you then continue your Cringes, when Old Age hath made her a Ghost, which only represents a hideous Picture of Death? In short, When you loved a Witty Man, it was because by the Quickness of his Apprehension, he unravelled an intricate Affair, seasoned the choicest Companies with his quaint Sayings, and sounded the depth of Sciences with a single Thought; and do you still honour him, when his worn Organs disappoint his weak Noddle, when he is become dull and uneasy in Company, and when he looks like an aged Fairy rather than a rational Man?

Conclude then from thence, Son, that it is fitter Young Men should govern Families, than Old ; and the rather, that according to your own Principles, *Hercules*, *Achilles*, *Epaminondas*, *Alexander*, and *Cæsar*, of whom most part died under Fourty Years of Age, could have merited no Honours, as being too Young in your account, though their Youth was the only cause of their Famous Actions ; which a more advanced Age would have rendered ineffectual, as wanting that Heat and Promptitude, that rendered them so highly successful. But you'll tell me, that all the Laws of your World, do carefully enjoin the Respect that is due to Old Men : That's true ; but it is as true also, that all who made Laws, have been Old Men, who feared that Young Men might justly have dispossessed them of the Authority they had usurped ———

You owe nothing to your mortal Architector, but your Body only ; your Soul comes from Heaven, and Chance might have made your Father your Son, as now you are his. Nay, are you sure he hath not hindered you from Inheriting a Crown ? Your Spirit left Heaven, perhaps with a design to animate the King of the *Romans*, in the Womb of the Emperess ; it casually encountered the *Embryo* of you

you by the way, and it may be to shorten its journey, went and lodged there: No, no, God would never have razed your name out of the List of Mankind, though your Father had died a Child. But who knows, whether you might not have been at this day the work of some valiant Captain, that would have associated you to his Glory, as well as to his Estate. So that, perhaps, you are no more indebted to your Father, for the life he hath given you, than you would be to a Pirate who had put you in Chains, because he feeds you: Nay, grant he had begot you a Prince, or King; a Present loses its merit, when it is made without the Option of him who receives it. *Cæsar* was killed, and so was *Cassius* too: In the mean time *Cassius* was obliged to the Slave, from whom he begg'd his Death, but so was not *Cæsar* to his Murderers, who forced it upon him. Did your Father consult your Will and Pleasure, when he Embraced your Mother? Did he ask you, if you thought fit to see that Age, or to wait for another; if you would be satisfied to be the Son of a Sot, or if you had the Ambition to spring from a Brave Man? Alas, you whom alone the business concerned, were

the

the only Person not consulted in the case. May be then, had you been shut up any where else, than in the Womb of Nature's Ideas, and had your Birth been in your own Opinion, you would have said to the *Parca*, my dear Lady, take another Spindle in your Hand: I have lain very long in the Bed of Nothing, and I had rather continue an Hundred years still without a Being, than to Be to day, that I may repent of it to morrow: However, Be you must, it was to no purpose for you to whimper and squall, to be back again, at the long and darksome House they drew you out of, they made as if they believed you cryed for the Teat.

These are the Reasons, at least some of them, my Son, why Parents bear so much respect to their Children: I know very well, that I have inclined to the Childrens side, more than in justice I ought; and that in favour of them, I have spoken a little against my Conscience. But since I was willing to repress the Pride of some Parents, who insult over the weakness of their little Ones; I have been forced to do as they do, who to make a crooked Tree streight, bend it to the contrary side, that betwixt two Conversions it may become

come even: Thus I have made Fathers reſtore to their Children what they have taken from them, by taking from them a great deal that belonged to them; that ſo another time they may be content with their own. I know very well alſo, that by this Apology I have offended all Old men: But let them remember, that they were Children before they were Fathers, and Young before they were Old; and that I muſt needs have ſpoken a great deal to their advantage, ſeeing they were not found in a Parſley-bed: But, in fine, fall back, fall edge, though my Enemies draw up againſt my Friends, it will go well enough ſtill with me; for I have obliged all men, and only diſobliged but one half.

With that he held his tongue, and our Landlord's Son ſpake in this manner: Give me leave, ſaid he to him, ſince by your care I am informed of the Original, Hiſtory, Cuſtoms, and Philoſophy of the World, of this little Man; to add ſomething to what you have ſaid; and to prove that Children are not obliged to Parents for their Generation, becauſe their Parents were obliged in Conſcience to procreate them.

The ſtricteſt Philoſophy of their World
ac-

acknowledges that, it is better to dye, since to dye, one muſt have lived, than not to have had a Being. Now ſeeing, by not giving a Being to that Nothing, I leave it in a ſtate worſe than Death, I am more guilty in not producing, than in killing it. In the mean time, my little Man, thou wouldſt think thou hadſt committed an unpardonable Parracide, ſhouldſt thou have cut thy Sons throat: It would indeed be an enormous Crime, but it is far more execrable, not to give a Being to that which is capable of receiving it: For that Child whom thou depriveſt of life for ever, hath had the ſatisfaction of having enjoyed it for ſome time. Beſides, we know that it is but deprived of it, but for ſome ages; but theſe forty poor little Nothings, which thou mighteſt have made forty good Souldiers for the King, thou art ſo malicious as to deny them Life, and letteſt them corrupt in thy Reins, to the danger of an Apoplexy, which will ſtifle thee.

This Philoſophy did not at all pleaſe me, which made me three or four times ſhake my Head; but our Preceptor held his tongue, becauſe Supper was mad to be gone.

We laid our selves along, then, upon very soft Quilts, covered with large Carpets; and a young man that waited on us, taking the oldest of our Philosophers, led him into a little arlour apart, where my Spirit called to him, to come back to us, assoon as he had supped.

This humour of eating separately, gave me the curiosity of asking the Cause of it: He'll not relish, said he, the steam of Meat, nor yet of Herbs, unless they die of themselves, because he thinks they are sensible of Pain. I wonder not so much, replied I, that he abstains from Flesh, and all things that have had a sensitive Life: For in our World the *Pythagoreans*, and even some holy *Anchorites*, have followed that Rule; but not to dare, for instance, cut a Cabbage, for fear of hurting it; that seems to me altogether ridiculous. And for my part, answered my Spirit, I find a great deal of probability in his Opinion.

For tell me, Is not that Cabbage you speak of, a Being existent in Nature, as well as you? Is not she the common Mother of you both? Nay it would even seem that she hath taken more care of the Vegetable, than of the Rational, since she hath referred the Generation of a Man a Man to the Caprices of his Father, who may, according as he has a mind to it, beget him, or not beget him: A rigour wherewith she hath not treated the Cabbage; for instead of remitting it to the discretion of the Father, to generate the Son, as if she had been more fearful, least the Race of Cabbage should perish, than that of Man; she forces them, whether they will or no, to give a Being to another, and not so as Men, who engender not, but according to their Whimseys; and who, during their whole life, cannot procreate above a Score; whereas Cabbages can produce many Thousands a Head. Yet the Opinion that Nature is kinder to Mankind, than to Cabbage-kind, tickles and makes us laugh: But seeing she is incapable of Passion, she can neither love, nor hate any thing; and were she susceptible of Love, she would rather bestow her affection upon this Cabbage, which you grant cannot offend her, than upon that Man who would destroy her, if it lay in his power. And moreover, Man cannot be born Innocent, being a Part of the first Offendor: But we know very well, that the first Cabbage did not offend its Creator. If it be said, that we are

are made after the Image of the Supreme Being, and so is not the Cabbage; grant that to be true; yet by polluting our Soul, wherein we resembled Him, we have effaced that Likeness, seeing nothing is more contrary to God than Sin. If then our Soul be no longer his Image, we resemble him no more in our Feet, Hands, Mouth, Forehead and Ears, than a Cabbage in its Leaves, Flowers, Stalk, Pith, and Head : Do not you really think, that if this poor Plant could speak, when one cuts it, it would not say, Dear Brother Man, what have I done to thee that deserves Death? I never grow but in Gardens, and am never to be found in desart places, where I might live in Security : I disdain all other company but thine; and scarcely am I sowed in thy Garden, when to shew thee my Goodwill, I blow, stretch out my Arms to thee; offer thee my Children in Grain; and as a requital for my civility, thou causest my Head to be chopt off. Thus would a Cabbage discourse, if it could speak. Well, and because it cannot complain, may we therefore justly do it all the Wrong which it cannot hinder? If I find a Wretch bound Hand and Foot, may I lawfully kill him, because he can-

not

not defend himſelf; ſo far from that, that his Weakneſs would aggravate my Cruelty? And though this wretched Creature be poor, and deſtitute of all the advantages which we have, yet it deſerves not Death; and when of all the Benefits of a Being, it hath only that of Encreaſe, we ought not cruelly to ſnatch that away from it. To maſſacre a Man, is not ſo great Sin, as to cut and kill a Cabbage, becauſe one day the Man will riſe again, but the Cabbage has no other Life to hope for: By putting to death a Cabbage, you annihilate it; but in killing a Man, you make him only change his Habitation: Nay, I'll go farther with you ſtill, ſince God doth equally cheriſh all his Works, and hath equally divided his Benefits betwixt Us, and Plants, it is but juſt we ſhould have an equal Eſteem for Them, as for our Selves. It is true, we were born firſt, but in the Family of God there is no Birth-right. If then the Cabbage ſhare not with us in the inheritance of Immortality, without doubt, that Want was made up by ſome other Advantage, that may make amends for the ſhortneſs of its Being; may be by an univerſal Intellect, or a perfect Knowledge of all things in their Cauſes;
and

and it's for that Reason, that the wise Mover of all things, hath not shaped for it Organs like ours, which are proper only for a simple Reasoning, not only weak, but many times fallacious too; but others, more ingeniously framed, stronger, and more numerous, which serve to manage its Speculative Exercises. You'll ask me, perhaps, when ever any Cabbage, imparted those lofty Conceptions to us? But tell me, again, who ever discovered to us certain Beings, which we allow to be above us; to whom, we bear no Analogy, nor Proportion, and whose Existence, it is as hard for us to comprehend, as the Understanding and Ways, whereby a Cabbage expresses its self to its like, though not to us, because our Senses are too dull to penetrate so far.

Moses, the greatest of Philosophers, who drew the Knowledge of Nature, from the Fountain-Head, Nature her self, hinted this truth to us, when he spoke of the Tree of Knowledge; and without doubt, he intended to intimate to us, under that Figure, that Plants, in Exclusion to Mankind, possess perfect Philosophy. Remember, then, O thou Proudest of Animals! that though a Cabbage which thou cuttest, sayeth not a Word, yet it pays it at Thinking; but the poor Vegetable, has no fit Or-

Organs to howl as you do, nor yet to frisk it about, and weep: Yet, it hath those that are proper to complain of the Wrong you do it, and to draw a Judgement from Heaven upon you for the Injustice. But if you still demand of me, how I come to know, that Cabbage and Coleworts conceive such pretty Thoughts? Then will I ask you, how come you to know that they do not? And that some amongst them, when they shut up at Night, may not Compliment one another, as you do, saying: Good Night, Master *Cole-Curled-Pate*; your most humble Servant, good Master *Cabbage-Round-Head*.

So far was he gone on in his Discourse, when the young Lad, who had led out our Philosopher, led him in again; What, Supped already? cryed my Spirit to him. He answered, yes, almost: The Physiognomist having permitted him to take a little more with us. Our young Landlord stayed not, till I should ask him the meaning of that Mystery; I perceive, said he, you wonder at this way of Living; know then, that in your World, the Government of Health is too much neglected, and that our Method is not to be despised.

In

In all Houses, there is a Physiognomist entertained by the Publick, who in some manner, resembles your Physicians, save that he only prescribes to the Healthful, and judges of the different manner, how we are to be Treated, only according to the Proportion, Figure, and Symetry of our Members; by the Features of the Face, the Complexion, the Softness of the Skin, the Agility of the Body, the Sound of the Voice, and the Colour, Strength, and Hardness of the Hair. Did not you just now mind a Man, of a pretty low Stature, why ey'd you: he was the Physiognomist of the House: Assure your self, that according as he observed your Constitution, he hath diversified the Exhalation of your Supper: Mark the Quilt on which you lie, how distant it is from our Couches; without doubt, he judged your Constitution to be far different from ours; since he feared, that the Odour which evaporates from those little Pipkins, that stand under our Noses, might reach you, or that yours might steam to us; at Night, you'll see him chuse the Flowers for your Bed, with the same Circumspection. During all this Discourse, I made Signs to my Landlord, that he would try, if he could oblige the Philosophers, to fall upon some head of the Science, which they professed. He was too much my Friend, not to start an Occasion upon the Spot: But not to trouble the Reader, with the Discourse, and Entreaties, that were previous to the Treaty, wherein Jest and Earnest were so wittily interwoven, that it can hardly be imitated; I'll only tell you, that the Doctor, who came last, after many things, spake as follows:

It remains to be proved, that there are infinite Worlds, in an infinite World: Fancy to your self then the Universe, as a great Animal, and that the Stars, which are Worlds, are in this great Animal, as other great Animals, that serve reciprocally for Worlds to other People. Such as we, our Horses, &c. That we in our turns, are likewise Worlds to certain other Animals, incomparably less than our selves, such as Nits, Lice, Hand-worms, &c. And that these are on Earth, to others more imperceptible ones; in the same manner, as every one of us, appears to be a great World, to these little People. Perhaps, our Flesh, Blood, and Spirits, are nothing else, but a Contexture of little Animals that correspond, lend us Motion from theirs, and blindly suffer themselves to be guided by our Will, which is their Coachman; or otherwise conduct us, and all Conspiring together, produce that Action which

which we call Life. For tell me, pray, is it a hard thing to be believed, that a Louse takes your Body for a World; and that when any one of them, travels from one of your Ears to the other, his Companions say, that he hath travelled the Earth, from end to end, or that he hath run from one Pole to the other? Yes, without doubt, those little People, take your Hair for the Forests of their Country; the Pores full of Liquor, for Fountains; Buboes and Pimples, for Lakes and Ponds; Boils, for Seas; and Defluxions, for Deluges: And when you Comb your self, forwards, and backwards, they take that Agitation, for the Flowing and Ebbing of the Ocean. Doth not Itching make good what I say? What is the little Worm that causes it, but one of these little Animals, which hath broken off from civil Society, that it may set up for a Tyrant in its Country? If you ask me, why are they bigger than other imperceptible Creatures? I ask you, why are Elephants bigger than we? And the *Irish*-men, than *Spaniards*? As to the Blisters, and Scurff, which you know not the Cause of; they must either happen by the Corruption of their Enemies, which these little Blades have killed, or which the Plague has caused by the scarcity of Food, for which the Seditious worried

ried one another, and left Mountains of Dead Carcases rotting in the Field; or because the Tyrant, having driven away on all Hands his Companions, who by their Bodies, stopt up the Pores of ours, hath made way out for the waterish Matter, which being extravasated, and out of the Sphere of the Circulation of our Blood, is corrupted. It may be asked, perhaps, why a Nit, or Hand-worm, produces so many disorders? But that's easily conceived; for as one Revolt begets another, so these little People, egg'd on by the bad Example of their Seditious Companions, aspire severally to Soveraign Command; and occasion every where, War, Slaughter, and Famine. But you'll say, some are far less subject to Itching, than others; and, nevertheless, all are equally inhabited by these little Animals, since you say, they are the Cause of our Life. That's true; for we observe, that Phlegmatick People, are not so much given to scratching as the Cholerick; because the People sympathizing with the Climate they inhabit, are slower in a cold Body, than those others that are heated by the temper of their Region, who frisk and stir, and cannot rest in a place: Thus a Cholerick Man, is more delicate than a Phlegmatick; because being animated in many more Parts, and the Soul, being the

Ai-

Action of these little Beasts, he is capable of Feeling, in all places where those Cattle stir. Whereas, the Phlegmatick Man, wanting sufficient Heat to put that stirring Mobile in Action, is sensible but in a few places; and to prove more plainly that universal *Vermicularity*, you need but consider, when you are wounded, how the Blood runs to the Sore: Your Doctors say, that it is guided by provident Nature, who would succour the parts debilitated; which might make us conclude, that, besides the Soul and Mind, there were a third intellectual Substance, that had distinct Organs and Functions: And therefore, it seems to me far more Rational, to say, That these little Animals, finding themselves attacked, send to demand Assistance from their Neighbours, and that Recruits flocking in from all Parts, and the Country being too little to contain so many, they either die of Hunger, or are stifled in the Press. That Mortality happens when the Boil is ripe; for as an Argument, that these Animals at that time are stifled, the Flesh becomes insensible: Now, if Blood-letting, which is many times ordered to divert the Fluxion, do any good, it is because, much being lost by the Orifice, which these little Animals laboured to stop, they refuse their Allies

Assistance, having no more Forces, than is enough to defend themselves at home.

Thus he concluded, and when the second Philosopher, perceived by all our Looks, that we longed to hear him speak in his turn:

Men, said he, seeing you are curious to instruct this little Animal, (our like) in somewhat of the Science which we profess, I am now dictating a Treatise, which I wish he might see, because of the Light it gives to the Understanding of our Natural Philosophy; it is an Explication of the Original of the World: But seeing I am in haste to set my Bellows at work, (for to Morrow, without delay, the Town departs;) I hope you'll excuse my want of time, and I promise to satisfie you, assoon as the Town is arrived, at the place whither it is to go.

At these words, the Landlord's Son called his Father, to know what it was a Clock? who having answered him, that it was past Eight, he asked him in a great Rage, Why he did not give him notice at Seven, according as he had commanded him, that he knew well enough, the Houses were to be gone to Morrow; and that the City Walls were already upon their Journey? Son, replyed the good Man, since you sate down to Table, there is an Order published, ed, That no House shall budg before next day: That's all one, answered the young Man; you ought blindly to obey, not to examine my Orders, and only remember what I commanded you. Quick, go fetch me your Effigies: So soon as it was brought, he took hold on't by the Arm, and Whipt it a whole quarter of an Hour: Away you ne'er be good, continued he, as a Punishment for your disobedience; it's my Will and Pleasure, that this day you serve for a Laughing-stock to all People; and therefore I command you, not to walk but upon two Legs, till Night. The Poor Man went out in a very mournful Condition, and the Young Man excused to us his Passion.

I had much ado, though I bit my Lip, to forbear Laughing at so pleasant a Punishment; and therefore to take me off of this odd piece of Pedantick Discipline, which, without doubt, would have made me burst out at last; I prayed my Philosopher to tell me, what he meant by that Journey of the Town, he talked of, and if the Houses and Walls Travelled? Dear Stranger, answered he, we have some Ambulatory Towns, and some Sedentary; the Ambulatory, as for instance, this wherein now we are, are Built in this manner: The Architector, as you see, builds every Palace of a very light sort of Timber; support-

ported by four Wheels underneath; in the thickness of one of the Walls, he places ten large pair of Bellows, whose Snouts pass in a Horizontal Line, through the upper Story, from one Pinacle to the other; so that when Towns are to be removed, from one place to another, (for according to the Seasons they change the Air) every one spreads a great many large Sails, upon one side of the House, before the Noses of the Bellows; then having wound up a Spring, to make them play, in less then Eight days time, their Houses by the continual Puffs, which these Windy Monsters blow, are driven, if one pleases, an Hundred Leagues and more. For those which we call Sendentary, they are almost like to your Towers; save that they are of Timber, and that they have a Great and Strong Skrew, or Vice in the Middle, reaching from the Top to the Bottom; whereby they may be hoisted up, or let down as People please. Now the Ground underneath, is dugg as deep as the House is high; and it is so ordered, that so soon as the Frosts begin to chill the Air, they may sink their Houses down under Ground, where they keep themselves secure from the Severity of the Weather: But assoon as the gentle Breathings of the Spring, begin to soften and qualifie the Air; they raise
them

them above Ground again, by means of the great Skrew I told you of. I prayed him, since he had shew'd me so much goodness, and that the Town was not to part till next day, that he would tell me somewhat of that Original of the World, which he had mentioned not long before; and I promise you, said I, that in requital, so soon as I am got back to the Moon, from whence my Governour (pointing to my Spirit) will tell you that I am come, I'll spread your Renown there, by relating the rare things you shall tell me: I perceive you Laugh at that promise, because you do not believe, that the Moon, I speak of, is a World, and that I am an Inhabitant of it; but I can assure you also, that the People of that World, who take this only for a Moon, will Laugh at me, when I tell them that your Moon is a World, and that there are Fields, and Inhabitants in it: He answered only with a smile, and spake in this manner:

Since in Ascending to the Original of this great *ALL*, we are forced to run into three or four Absurdities; it is but reasonable we should follow the way, wherein we may be least apt to stumble. I say then, that the first Obstacle that stops us short, is the Eternity of the World; and the minds of men, not being able enough to con-

conceive it, and being no more able to imagine, that this great Universe, so lovely and so well ordered, could have made it self, they have had their recourse to Creation: But like to him that would leap into a River, for fear of being wet with Rain, they save themselves out of the Clutches of a Dwarf, by running into the Arms of a Giant; and yet they are not safe for all that: For that Eternity which they deny the World, because they cannot comprehend it, they attribute it to God, as if he stood in need of that Present, and as if it were easier to imagine it, in the one than in the other; for tell me, pray, was it ever yet conceived in Nature, how Something can be made of Nothing? Alas! betwixt Nothing and an Atome only, there are such infinite Disproportions, that the sharpest Wit could never dive into them; therefore to get out of this inextricable Labyrinth, you must admit of a Matter Eternal with God: But you'l say to me, grant I should allow you that Eternal Matter; how could that Chaos dispose and order it self? That's the thing I am about to explain to you.

My little Animal, after you have mentally divided every little Visible Body, into an infinite many little invisible Bodies; you must imagine, That the infinite Universe
con-

consists only of these Atomes, which are most solid, most incorruptible, and most simple; whose Figures are partly Cubical, partly Parallelograms, partly Angular, partly Round, partly Sharp-pointed, partly Pyramidal, partly Six-cornered, and partly Oval; which act all severally, according to their Various Figures: And to shew that it is so, put a very round Ivory Bowl upon a very smooth place, and with the least touch you give, it will be half a quarter of an hour before it rest: Now I say, that if it were perfectly round, as some of the Atomes I speak of are, and the Surface on which it is put perfectly smooth, it would never rest. If Art then be capable of inclining a Body to a perpetual Motion, why may we not believe that Nature can do it? It's the same with the other Figures, of which the Square require a perpetual Rest, others an oblique Motion, others a half Motion, as Trepidation; and the Round, whose Nature is to move, joyning a Pyramidal, makes that, perhaps, which we call Fire; because not only Fire is in continual Agitation, but also because it easily penetrates: Besides, the Fire hath different effects, according to the opening and quality of the Angles, when the round Figure is joyned; for Example, The Fire of Pepper is another thing, than the Fire of Sugar,

Sugar, the Fire of Sugar differs from that of Cinnamon; that of Cinnamon, from that of the Clove; and this from the Fire of a Faggot. Now the Fire, which is the Architect of the parts and whole of the Universe, hath driven together, and Congregated into an Oak, the quantity of Figures which are necessary for the Composition of that Oak: But you'l say, how could Hazard congregate into one place, all the Figures that are necessary for the production of that Oak? I answer, That it is no wonder that Matter so disposed, should form an Oak; but the wonder would have been greater, if the Matter being so disposed, the Oak had not been produced, had there been a few less of some Figures, it would have been an Elm, a Poplar, a Willow; and fewer of 'em still, it would have been the Sensitive Plant, an Oyster, a Worm, a Flie, a Frog, a Sparrow, an Ape, a Man. If three Dice being flung upon a Table, there happen a Raffle of two, or all, a three, a four, and a five, or two sixes, and a third in the bottom, would you say, O strange! that each Die should turn up such a chance, when there were so many others. A Sequence of three hath happened, O strange! Two sixes turned up, and the bottom of the third, O strange! I am sure that being a man of Sense, you'l never

make

World of the Moon.

make such Exclamations; for since there is but a certain quantity of Numbers upon the Dice, it's impossible but some of them must turn up; and you wonder, after that, how matter shuffled together *Pell-Mell*, as Chance pleases, should make a Man, seeing so many things were necessary for the Construction of his Being. You know not then, that this Matter tending to the Fabrick of a Man, hath been a Million of times stopt in it's Progress, for forming sometimes a Stone, sometimes Lead, sometimes Coral, sometimes a Flower, sometimes a Comet; and all because of more or less Figures, that were required for the framing of a Man: So that it is no greater wonder, if amongst infinite Matters, which incessantly change and stir, some have hit upon the construction of the few Animals, Vegetables, and Minerals which we see, than if in a Hundred Casts of the Dice, one should throw a Raffle: Nay, indeed, it is impossible, that in this hurling of things, nothing should be produced; and yet this will be always admired by a Block-head, who little knows how small a matter, would have made it to have been o-

herwise. When the great River of
.akes a Mill to Grind, guides the Wheels
f a Clock, and the Brook of
only

only runs, and sometimes absconds, you will not say, that that River hath a great deal of Wit, because you know that it hath met with things, disposed for producing such rare Feats; for had not the Mill stood in the way, it would not have ground the Corn; had it not met the Clock, it would not have marked the Hours: and if the little Rivulet I speak of, had met with the same Opportunities, it would have wrought the very same Miracles. Just so it is with the Fire that moves of it self; for finding Organs fit for the Act of Reasoning, it Reasons; when it finds only such, as are proper for Sensation, it Sensates; and when such as are fit for Vegetation, it Vegetates. And to prove it is so, put out but the Eyes of a Man, the Fire of whose Soul makes him to see, and he will cease to see; just as our great Clock will leave off to make the Hours, if the Movements of it be broken.

In fine, these Primary and indivisible Atomes, make a Circle, whereon without difficulty move the most perplexed Difficulties of Natural Philosophy; not so much as even the very Operation of the Senses, which no Body hitherto hath been able to conceive, but I will easily explain by these little Bodies. Let us begin with the Sight. It deserves, as being the most incomprehensible, our first Essay.

It is performed then, as I imagine, when the Tunicles of the Eye, whose Pores resemble those of Glass, transmitting that fiery Dust, which is called Visual Rays, the same is stopt by some opacous Matter, which makes it recoil; and then, meeting in its retreat the Image of the Object, that forced it back, and that Image, being but an infinite number of little Bodies, exhaled in an equal Superfice, from the Object beheld; it pursues it to our Eye: You'll not fail to Object, I know, that Glass is an Opacous Body, and very Compact; and that neverthelefs, instead of reflecting other Bodies, it lets them pass through: But I answer, that the Pores of Glass, are shaped in the same Figure, as those Atomes are which pass through it; and as a Wheat-Sieve is not proper for Sifting of Oats, nor an Oat-Sieve to Sift Wheat; so a Box of Deal-Board, though it be thin, and lets a sound go through it, is impenetrable to the Sight; and a piece of Chrystal, though transparent, and pervious to the Eye, is not penetrable to the Touch. I could not here forbear to interrupt him: A great Poet and Philosopher of our World, said I, hath, after *Epicurus* and *Democritus*, spoken of these little Bodies, in the same manner almost as you do; and therefore, you don't at all surprise me, by

that

that Discourse: Only, tell me, I pray, as you proceed, how, according to your Principles, you'll explain to me, the manner of drawing your Picture in a Looking-Glass. That's very easie, replied he, for imagine with your self, that those Fires of our Eyes, having passed through the Glass, and meeting behind it an Opacous Body, that reverberates them, they come back the way they went; and finding those little Bodies, marching in equal Superfices upon the Glass, they repel them to our Eyes; and our Imagination, hotter than the other Faculties of our Soul, attracts the more subtile, wherewith it draws our Picture in little.

It is as easie to conceive the Act of Hearing, and for *Brevities* sake, let us only consider it in the Harmony of a Lute, touched by the Hand of a Master. You'll ask me, How can it be, that I perceive at so great a distance, a thing which I do not see? Does there a Sponge go out of my Ears, that drinks up that Musick, and brings it back with it again? Or does the Player, beget in my Head another little Musician, with another little Lute, who has Orders like an Eccho, to sing over to me the same Airs? No; But that Miracle proceeds from this, that the String touched, striking those little Bodies, of which the
Air

Air is composed, drives it gently into my Brain, with those little Corporeal Nothings, that sweetly pierce into it; and according as the String is stretched, the Sound is high, because it more vigorously drives the Atomes; and the Organ being thus penetrated, furnisheth the Fancy, wherewith to make a Representation; if too little, then our Memory, not having as yet finished its Image, we are forced to repeat the same sound to it again; to the end it may take enough of Materials, which, for Instance, the Measures of a *Saraband*, furnish it with, for finishing the Picture of that *Saraband*; but that Operation, is nothing near so wonderful, as those others, which by the help of the same Organ, excite us sometimes to Joy, sometimes to Anger.——

And this happens, when in that motion, these little Bodies, meet with other *little* Bodies within us, moving in the same manner, or whose Figure, renders them susceptible of the same Agitation; for then these New-comers, stir up their Landlords to move as they do; & thus, when a violent Air meets with the Fire of our Blood, it inclines it to the same Motion, and animates it to a Sally, which is the thing we call Heat of Courage; if the Sound be softer, and have only force enough to raise a less Flame, in greater Agitation, by leading it along the Nerves,

Membranes, and through the interstices of our Flesh, it excites that Tickling which is called Joy: And so it happens, in the Ebullition of the other Passions, according as these little Bodies, are more or less violently tossed upon us, according to the Motion they receive, by the rencounter of other Agitations, and according as they find Dispositions in us for motion. So much for Hearing.

Now, I think the Demonstration of Touching, will be every whit as easie; if we conceive, that out of all palpable Matter, there is a perpetual Emission of little Bodies, and that the more we touch them, the more evaporate; because, we press them out of the Subject it self, as Water out of a Sponge, when we squeez it. The Hard, make a report to the Organ, of their Hardness; the Soft, of their Softness; the Rough, &c. And that this is so, we are not so quaint in Feeling with Hands used to Labour, because of the Thickness of the Skin, which being neither porous, nor animated, with difficulty transmits the Evaporations of Matter. Some, perhaps, may desire to know, where the Organ of Touching has its Residence. For my part, I think it is spread over all the Surface of the Body, seeing in all parts it feels: Yet I imagine, that the nearer the Member, wherewith

with we touch, be to the Head, the sooner we distinguish; which Experience convinces us of, when with shut Eyes we handle any thing, for then we'll more easily guess what it is; and if on the contrary, we feel it with our hinder Feet, it will be harder for us to know it: And the Reason is, because our Skin being all over perforated, our Nerves, which are of no compacter Matter, lose by the way a great many of those little Atomes, through the little Holes of their Contexture, before they reach the Brain, which is their Journeys end: It remains, that I speak of the Smelling and Tasting.

Pray tell me, when I taste a Fruit, is it not because the Heat of my Mouth melts it? Confess to me then, that there being Salts in a Pear, and that they being separated by Dissolution, into little Bodies of a different Figure, from those which make the Taste of an Apple; they must needs pierce our Pallate, in a very different manner: Just so as the thrust of a Pike, that passes through me, is not like the Wound which a Pistol-Bullet makes me feel, with a sudden start; and as that Pistol-Bullet, makes me suffer another sort of Pain, than that of a Slug of Steel.

I have nothing to say, as to the Smelling, seeing the Philosophers themselves confess, that it is performed by a continual Emission of little Bodies.

Now, upon the same Principle, will I explain to you, the Creation, Harmony, and Influence of the Celestial Globes, with the immutable Variety of Meteors.

He was about to proceed; but the Old Landlord coming in, made our Philosopher think of withdrawing: He brought in Christals full of Glow-worms, to light the Parlour; but seeing those little fiery Insects, lose much of their Light, when they are not fresh gathered, these which were ten days old, had hardly any at all. My Spirit stayed not, till the Company should complain of it, but went up to his Chamber, and came immediately back again, with two Bowls of Fire so Sparkling, that all wondred he burnt not his Fingers. These incombustible Tapers, said he, will serve us better than your Week of Worms. They are Rays of the Sun, which I have purged from their Heat; otherwise, the corrosive qualities of their Fire, would have dazled, and offended your Eyes; I have fixed their Light, and inclosed it within these transparent Bowls. That ought not to afford you any great Cause of Admimiration; for it is not harder for me, who am a Native of the Sun, to condense his Beams, which are the Dust of that World, than it is for you, to gather the Atomes of the pulveriz'd Earth of this World. Thereupon our Landlord sent a Servant, to wait upon the Philosophers home, it being then Night, with a dozen Globes of Glowworms hanging at his four Legs. As for my Preceptor, and my self, we went to rest, by order of the Phisiognomist. He laid me that Night in a Chamber of Violets and Lillies, ordered me to be tickled after the usual manner; and next Morning about Nine a Clock, my Spirit came in, and told me, that he was come from Court, where One of the Queens Maids of Honour had sent for him, and that she had enquired after me, protesting that she still persisted in her Design, to be as good as her Word; that is, that with all her Heart she would follow me, if I would take her along with me to the other World; which exceedingly pleased me, said he, when I understood, that the chief Motive, which inclined her to the Voyage, was to become Christian: And therefore, I have promised to forward her Design, what lies in me; and for that end, to invent a Machine, that may hold three or four, wherein you may mount to day, both together, if you think fit. I'll go seriously set about the performance of my Undertaking; and in the mean time, to entertain you, during my Absence, I leave you here a Book, which heretofore I brought with me from my Native Countrey; the Title of it is, *The States and Empires of the Sun, with an Addition of the History of the Spark*. I also give you this, which I esteem much more, it is the great Work of the Philosophers, composed by one of the greatest Wits of the Sun. He proves in it, that all things are true, and shews the way of uniting Physically, the Truths of every Contradiction; as, for Example, That White is Black, and Black White; that one may be, and not be at the same time; that there may be a Mountain without a Valley; that nothing is something, and that all things that are, are not; but observe, that he proves all these unheard-of Paradoxes, without any Captious or Sophistical Argument. When you are weary of Reading, you may Walk, or Converse with our Landlord's Son, he has a very Charming Wit; but that which I dislike in him is, that he is a little Atheistical. If he chance to Scandalize you, or by any Argument shake your Faith, fail not immediately to come and propose it to me, and I'll clear the Difficulties of it; any

other

other, but I, would enjoin you to break Company with him; but since he is extreamly proud and conceited, I am certian, he would take your flight for a Defeat, and would believe your Faith to be grounded on no Reason, if you refused to hear his. Having said so, he left me; and no sooner was his back turned, but I fell to consider attentively my Books and their Boxes, that's to say, their Covers, which seemed to me to be wonderfully Rich; the one was cut of a single Diamond, incomparably more resplendent than ours; the second looked like a prodigious great Pearl, cloven in two. My Spirit had translated those Books into the Language of that World; but because I have none of their Print, I'll now explain to you the Fashion of these two Volumes.

As I opened the Box, *I* found within somewhat of Metal, almost like to our Clocks, full of I know not what little Springs, and imperceptible Engines: It was a Book, indeed; but a Strange and Wonderful Book, that had neither Leaves nor Letters: In fine, it was a Book, made wholly for the Ears, and not the Eyes. So that when any Body has a mind to read in it, he winds up that Machine, with a great many little Strings; then he turns the Hand to the Chapter which he desires to hear, and

and straight, as from the Mouth of a Man, or a Musical Instrument, proceed all the distinct and different Sounds, which the *Lunar* Grandees make use of, for expressing their Thoughts, instead of Language.

When I since reflected on this Miraculous Invention, I no longer wondred, that the Young-Men of that Country, were more knowing at Sixteen, or Eighteen years Old, than the Gray-Beards of our Climate; for knowing how to Read as soon as Speak, they are never without Lectures, in their Chambers, their Walks, the Town, or Travelling; they may have in their Pockets, or at their Girdles, Thirty of these Books, where they need but wind up a Spring, to hear a whole Chapter, and so more, if they have a mind to hear the Book quite through; so that you never want the Company of all the great Men, Living and Dead, who entertain you with Living Voices. This Present employed me about an hour; and then hanging them to my Ears, like a pair of Pendants, I went a Walking; but I was hardly at End of the Street, when I met a Multitude of People very Melancholy.

Four of them carried upon their Shoulders, a kind of a Herse, covered with Black: I asked a Spectator, what that Procession

cession, like to a Funeral in my Country, meant? He made me answer, that that naughty ▬▬ called so by the People, because of a knock he had received upon the Right Knee; who being convicted of Envy and Ingratitude, died the day before; and that Twenty Years ago, the Parliament had Condemned him to die in his Bed, and then to be interred after his Death. I fell a Laughing at that Answer. And he asking me, why? You amaze me, said I, that that which is counted a Blessing in our World, as a long Life, a peaceable Death, and an Honourable Burial, should pass here for an exemplary Punishment. What, do you take a Burial, for a precious thing then, replyed that Man? And, in good earnest, can you conceive any thing more Horrid, than a Corps crawling with Worms, at the discretion of Toads, which feed on his Cheeks; the Plague it self Clothed with the Body of a Man? Good God! The very thought of having, even when I am Dead, my Face wrapt up in a Shroud, and a Pike-depth of Earth upon my Mouth, makes me I can hardly fetch breath. The Wretch whom you see carried here, besides the disgrace of being thrown into a Pit, hath been Condemned, to be attended by an Hundred and Fifty

of

of his Friends; who are strictly charged, as a Punishment for their having loved an envious and ungrateful Person, to appear with a sad Countenance at his Funeral; and had it not been that the Judges took some compassion of him, imputing his Crimes partly to his want of Wit, they would have been commanded to Weep there also. All are Burnt here, except Malefactors: And, indeed, it is a most rational and decent Custom: For we believe, that the Fire having separated the pure from the impure, the Heat by Sympathy reassembles the natural Heat; which made the Soul, and gives it force to mount up, till it arrive at some Star, the Country of certain people, more immaterial and intellectual than us; because their Temper ought to suit with, and participate of the Globe which they inhabit.

However, this is not our neatest way of Burying neither; for when any one of our Philosophers comes to an Age, wherein he finds his Wit begin to decay, and the Ice of his years, to numm the Motions of his Soul, he invites all his Friends to a sumptuous Banquet; then having declared to them the Reasons, that move him to bid farewel to Nature, and the little hopes he has, of adding any thing more to his worthy Actions, they shew him Favour;
that's

that's to say, they suffer him to Dye, or otherwise are severe to him, and command him to Live. When then, by plurality of Voices, they have put his Life into his own Hands, he acquaints his dearest Friends with the day and place. These purge, and for Four and Twenty hours abstain from Eating; then being come to the House of the Sage, and having Sacrificed to the Sun, they enter the Chamber, where the generous Philosopher waits for them on a Bed of State; every one embraces him, and when it comes to his turn, whom he loves best, having kissed him affectionately, leaning upon his Bosom, and joyning Mouth to Mouth, with his right hand he sheaths a Dagger in his Heart. The Loving Friend parts not his Lips from his Friends Lips, till he find him expired; and then pulling out the Steel, and putting his Mouth close to the Wound, he sucks down his Blood, till a Second succeed him, then a Third, Fourth, and so all the Company: Four or Five Hours after, every one has a Young Wench, of Sixteen or Seventeen Years of Age, brought to him; and during Three or Four days, whilst they are tasting the Pleasures of Love, they feed on nothing but the Flesh of the Deceased, which they eat raw; to the end that if from an Hundred Embracements any thing Spring,

they

they may be assured it is their old Friend Revived.

I interrupted this Discourse, saying to him that told me all, That this Manner of Acting much resembled the ways of some People of our World; and so pursued my Walk, which was so long, that when I came back, Dinner had been ready Two Hours. They asked me, why I came so late? It is not my Fault, said I to the Cook, who complained: I asked what it was a Clock several times in the Street, but they made me no answer, but by opening their Mouths, shutting their Teeth, and turning their Faces awry.

How, cried all the Company, did not you know by that, that they shewed you what it was a Clock? Faith, said I, they might have held their great Noses in the Sun long enough, before I had understood what they meant. It's a Commodity, said they, that saves them the Trouble of a Watch; for with their Teeth they make so true a Dial, that when they would tell any Body the Hour of the day, they do no more but open their Lips, and the shadow of that Nose, falling upon their Teeth, like the Gnomon of a Sun-Dial, makes the precise time. Now that you may know the reason, why all People in this Country have great Noses; assoon as a Woman is

brought

brought to Bed, the Midwife carries the Child to the *Master of the Seminary*; and exactly at the years end, the Skillful being assembled, if his Nose prove shorter than the standing Measure, which an Alderman keeps, he is judged to be a *Flat Nose*, and delivered over to be gelt. You'l ask me, no doubt, the Reason of that Barbarous Custom, and how it comes to pass, that we, amongst whom Virginity is a Crime, should enjoyn Continence by force; but know, that we do so, because after Thirty Ages experience we have observed, that a great Nose is the mark of a Witty, Courteous, Affable, Generous and Liberal Man; and that a little Nose is a Sign of the contrary: Wherefore of *Flat Noses* we make Eunuchs, because the Republick had rather have no Children at all, than Children like them. He was still a speaking, when I saw a man come in stark Naked; I presently sat down and put on my Hat to shew him Honour, for these are the greatest Marks of Respect, that can be shew'd to any in that Country. The Kingdom, said he, desires you would give the Magistrates notice, before you return to your own World; because a Mathematician hath just now undertaken before the Council, that provided when you are returned home, you would make a certain Machine, that he'l teach you how

to

to do; he'l attract your Globe, and joyn it to this. Good now, (said I to my Landlord, when the other was gone) tell me why that Messenger carried at his Girdle, Privy Members of Brass; a thing I have often seen, whilst I was in my Cage, but durst not ask the Reason, because I was always environed by the Queens Maids of Honour, whom I feared to offend, if in their presence I had talked of such a foul Subject? He made me this answer: The Females here, no more than the Males, are not so ungrateful, as to blush at the sight of that which Forged them; and Virgins are not ashamed to love upon us, in Memory of Mother Nature, the only thing that represents her best. Know then, that the Scarf wherewith that Man is Honoured, and which for a Medal has the Bauble of a Man hanging at it, is the Badg of a Gentleman, and the Mark to distinguish the Cavalier from the Clown. This seemed to me, to be so extravagant a Paradox, that I could not forbear Laughing. I look upon that, replyed I, to be a very extraordinary Custom, for in our World, to wear a Sword is the Badg of a Gentleman. But, my dear little Man, cried my Host without startling, what, are the great Men of your World Mad then; to make ostentation

tation of an Instrument, that's the mark of a Hang man, made only to destroy us; and in a word, the sworn Enemy of all that has Life? And on the contrary, to hide a Member, without which, we had been ranked among the things that are not; the *Prometheus* of every Animal, and the indefatigable Repairer of the Frailties of Nature? Unhappy Country! where the Marks of Generation are Ignominious, and those of Destruction Honourable: In the mean time, you call that Member the shameful Privy-Parts, as if any thing were more Glorious, than to give Life, or any thing more disgraceful, than to take it away. During all this Discourse, we went on with our Dinner; and as soon as we rose from Table, we went to take the Air in the Garden; where taking Occasion to speak of the Generation, and Conception of things, he said to me, You must know, that the Earth, converting it self into a Tree, from a Tree into a Hog, and from a Hog into a Man, is an Argument, that all things in Nature, aspire to be Men; since that is the most perfect Being, as being a Quintessence, and the best devised Mixture in the World; which alone unites the Animal and Rational Life into one. None but a Pedant will deny me this, when we see that a Plumb-Tree, by the Heat of its Germ,

K as

as by a Mouth, sucks in and digests the Earth that's about it; that a Hog devours the Fruit of this Tree, and converts it into the Substance of it self; and that a Man feeding on that Hog, reconcocts that dead Flesh, unites it to himself, and makes that Animal to revive under a more Noble Species. So the Man whom you see, perhaps threescore years ago, was no more but a Tuft of Grass in my Garden; which is the more probable, that the Opinion of the *Pythagorean Metamorphosis*, which so many Great Men maintain, in all likelyhood has only reached us, to engage us into an Enquiry after the truth of it; as, in reality, we have found that Matter, and all that has a Vegetative or Sensitive Life, when once it hath attained to the period of its Perfection, wheels about again, and descends into its Inanity, that it may return upon the Stage, and Act the same Parts over and over. I went down extreamly satisfyed to the Garden, and was beginning to rehearse to my Companion, what our Master had taught me; when the Physiognomist came to conduct us to Supper, and afterwards to Rest.

Next Morning, so soon as I awoke, I went to call up my Antagonist. It is, said I, accosting him, as great a Miracle to find a great Wit, like yours, buried in Sleep, as to see Fire, without Heat and Action. He bore with this ugly Compliment; but, (cryed he, with a Cholerick kind of Love) will you never leave these Fabulous Terms? Know, that these Names defame the Name of a Philosopher; and that seeing the wise Man sees nothing in the World, but what he conceives, and judges may be conceived, he ought to abhor all those Expressions of Prodigies, and extraordinary Events of Nature, which Block-heads have invented, to excuse the Weakness of their Understanding.

I thought my self then obliged in Conscience, to endeavour to undeceive him; and therefore, said I, though you be very stiff and obstinate in your Opinions, yet I have plainly seen supernatural Things happen. Say you so, continued he; you little know, that the force of Imagination, is able to cure all the Diseases, which you attribute to supernatural Causes, by reason of a certain natural Balsam, that contains Qualities quite contrary to the qualities of the Diseases that attack us; which happens, when our Imagination informed by Pain, searches in that place for the specifick Remedy, which it applies to the Poison. That's the reason, why an able Physician in your World, advises the Patient to make use of an Ignorant Doctor, whom he esteems to be very knowing, rather than of a very Skilful Physician, whom he may imagine to be Ignorant; because he fancies, that our Imagination labouring to recover our Health, provided it be assisted by Remedies, is able to cure us; but that the strongest Medicines are too weak, when not applied by Imagination. Do you think it strange, that the first Men of your World lived so many Ages, without the least Knowledge of Physick? No. And what might have been the Cause of that, in your judgement; unless, their Nature was as yet in its force, and that natural Balsam in vigour, before they were spoilt by the Drugs, wherewith Physicians consume you; it being enough then for the recovery of ones Health, earnestly to wish for it, and to imagine himself cured? So that their vigorous Fancies, plunging into that vital Oyl, extracted the Elixir of it, and applying Actives to Passives, in almost the twinkling of an Eye, they found themselves as sound as before: Which, notwithstanding the Depravation of Nature, happens even at this day, though somewhat rarely; and is by the Multitude called a Miracle: For my part, I believe not a jot on't, and have this to say for my self, that it is easier for all these Doctors to be mistaken, than that the other may not easily come to pass: For I put the Question to them

them; A Patient recovered out of a Feaver, heartily defired, during his ficknefs, as it is like, that he might be cured, and, may be, made Vows for that effect; fo that of neceffity he muft either have dyed, continued fick, or recovered: Had he died, then would it have been faid, kind Heaven hath put an end to his Pains; Nay, and that according to his Prayers, he was now cured of all Difeafes, praifed be the Lord: Had his Sicknefs continued, one would have faid, he wanted Faith; but becaufe he is cured, it's a Miracle forfooth. Is it not far more likely, that his Fancy being excited by violent Defires, hath done its Duty, and wrought the Cure? For grant he hath efcaped, what then? muft it needs be a Miracle? How many have we feen, pray, and after many folemn Vows and Proteftations, go to pot with all their fair Promifes and Refolutions.

But, at leaft, replied I, to him, if what you fay of that Balfam be true, it is a mark of the Rationality of our Soul; feeing without the help of our Reafon, or the Concurrence of our Will, fhe Acts of her felf; as if being without us, fhe applied the Active to the Paffive. Now, if being feparated from us fhe is Rational, it neceffarily follows, that fhe is Spiritual; and if you acknowledge her to be Spiritual, I conclude fhe is immortal; seeing Death happens to Animals, only by the changing of Forms, of which Matter alone is capable. The Young Man at that, decently sitting down upon his Bed, and making me also to sit, discoursed, as I remember, in this manner: As for the Soul of Beasts, which is Corporeal, I do not wonder they Die; seeing the best Harmony of the four Qualities may be dissolved, the greatest force of Blood quelled, and the loveliest Proportion of Organs disconcerted; but I wonder very much, that our intellectual, incorporeal, and immortal Soul, should be constrained to dislodge, and leave us by the same Cause, that makes an Ox to perish. Hath she covenanted with our Body, that as soon as he should receive a prick with a Sword in the Heart, a Bullet in the Brain, or a Musket-shot through the Chest, she should pack up and be gone; ——and if that Soul were Spiritual, and of her self so Rational, that being separated from our Mass, she understood as well as when Clothed with a Body; why cannot Blind Men, born with all the fair advantages of that intellectual Soul, imagine what it is to see? Is it, because they are not as yet deprived of Sight, by the Death of all their Senses? How! I cannot then make use of my Right Hand, because I have a Left.——And in fine, to make a just com-

comparison, which will overthrow all that you have said; I shall only alledge to you a Painter, who cannot work without his Pencil: And I'll tell you, that it is just so with the Soul, when she wants the use of the Senses. O yes, but added he——In the mean time, they'l have this Soul, which can only act imperfectly, because of the loss of one of her Tools, in the course of Life, to be able then to work to Perfection, when after our death, she hath lost them all. If they tell me, over and over again, that she needeth not these Instruments, for performing her Functions, I'll tell them e'en so, That then all the Blind about the Streets, ought to be Whipt at a Carts-Arse, for playing the Counterfeits, in pretending not to See a bit. He would have gone on in such impertinent Arguments, had not I stopt his Mouth, by desiring him to forbear, as he did for fear of a quarrel; for he perceived I began to be in a heat: So that he departed, and left me admiring the People of that World, amongst whom even the meanest have Naturally so much Wit; whereas those of ours have so little, and yet so dearly bought. At length my Love for my Country, took me off of the desire and thoughts I had of staying there, I minded nothing now but to be gone; but I saw so much impossibi-
lity

lity in the matter, that it made me quite peevish and melancholick. My Spirit observed it, and having asked me, What was the reason, that my Humor was so much altered? I frankly told him the Cause of my Melancholy; but he made me such fair Promises concerning my Return, that I relied wholly upon him. I acquainted the Council with my design; who sent for me, and made me take an Oath, that I should relate in our World, all that I had seen in that. My Pass-ports then were expeded, and my Spirit having made necessary Provisions for so long a Voyage, asked me, What part of my Country I desired to light in? I told him, that since most of the Rich Youths of *Paris*, once in their life time, made a Journey to *Rome*; imagining after that, that there remained no more worth the doing or seeing, I prayed him to be so good as to let me imitate them: But withal, said I, in what Machine shall we perform the Voyage, and what Orders do you think the Mathematician, who talked t'other day of joyning this Globe to ours, will give me? As to the Mathematician, said he, let that be no hinderance to you; for he is a Man who promises much, and performs little or nothing. And as to the Machine, that's to carry you back, it shall be the same which brought you to Court. How, said I, will

the

the Air become as solid as the Earth, to bear your steps? I cannot believe that: And it is strange, replied he, that you should believe, and not believe. Pray why should the Witches of your World, who march in the Air, and conduct whole Armies of Hail, Snow, Rain, and other Meteors, from one Province into another, have more Power than we? Pray have a little better opinion of me, than to think I would impose upon you. The truth is, said I, I have received so many good Offices from you, as well as *Socrates*, and the rest, for whom you have so great kindness, that I dare trust my self in your hands, as now I do, resigning my self heartily up to you. I had no sooner said the word, but he rose like a Whirl-wind, and holding me between his Arms, without the least Uneasiness, he made me pass that vast space, which Astronomers reckon betwixt the Moon and us, in a day and a halfs time; which convinced me, that they tell a Lye, who say that a Mill-stone would be Three Hundred Threescore, and I know not how many years more, in falling from Heaven, since I was so short a while, in dropping down from the Globe of the Moon upon this. At length, about the beginning of the Second day, I perceived I

was

was drawing near our World; since I could already distinguish *Europe* from *Africa*, and both from *Asia*, when I smelt Brimstone, which I saw steaming out of a very high Mountain, that incommoded me so much that I fainted away upon it. I cannot tell what befel me afterwards; but coming to my self again, I found I was amongst Briers on the side of a Hill, amidst some Shepherds, who spake *Italian*. I knew not what was become of my Spirit, and I asked the Shepherds if they had not seen him. At that word they made the sign of the Cross, and looked upon me, as if I had been a Devil my self: But when I told them that I was a Christian, and that I begg'd the Charity of them, that they would lead me to some place, where I might take a little rest; they conducted me into a Village, about a Mile off; where no sooner was I come, but all the Dogs of the place, from the least Cur, to the biggest Mastiff, flew upon me, and had torn me to pieces, if I had not found a House, wherein I saved my self: But that hindered them not to continue their Barking and Bawling, so that the Master of the House began to look upon me with an Evil Eye; and really I think, as people are very apprehensive, when Accidents which they look upon to be ominous happen, that man could have de-

delivered me up, as a Prey to these accursed Beasts, had not I bethought my self, that that which madded them so much at me, was the World from whence I came; because being accustomed to bark at the Moon, they smelt I was come from thence, by the scent of my Cloaths, which stuck to me, as a Sea-smell hangs about those, who have been long on Ship-board, for sometime after they come ashore. To Air my self then, I lay three or four hours in the Sun, upon a Terrass-walk; and being afterwards come down, the Dogs, who smelt no more that influence which had made me their Enemy, left barking, and peaceably went to their several homes. Next day I parted for *Rome*, where I saw the ruins of the Triumphs of some great Men, as well as of Ages: I admired those lovely Relicks; and the Repairs of some of them made by the Modern. At length, having stayed there a fortnight in Company of *Monsieur de Cyrano* my Cousin, who advanced me Money for my Return, I went to *Civita vecchia*, and embarked in a Galley that carried me to *Marseilles*. During all this Voyage, my mind run upon nothing, but the Wonders of the last I made. At that time I began the Memoires of it; and after my return, put them into as good order, as Sickness, which confines me to Bed, would permit. But foreseeing, that it will put

put an end to all my Studies, and Travels; that I may be as good as my word to the Council of that World; I have begg'd of *Monsieur le Bret*, my dearest and most constant Friend, that he would publish them, with the History of the *Republick of the Sun*, that of the *Spark*, and some other Pieces of my Composing, if those who have Stolen them from us, restore them to him, as I earnestly adjure them to do.

FINIS.

ERRATA.

PAge 17. line *ult.* read *Telescope.* p. 39. l. 18. add *long.* p. 58. l. 5. r. were. p. 65. l. 2. r. ends. p. 99. l. 14. r. who. p. 100. l. 21. r. an.

さい

THE
Comical History,
OF THE
STATES and EMPIRES
OF THE
WORLD
OF THE
SUN.

Written in *French* by *Cyrano Bergerac*.

And now *Englished* by *A. Lovell.* A. M.

LONDON,
Printed for *Henry Rhodes*, next door
Swan-Tavern, near *Bride-Lane* in
Street, 1687.

THE
HISTORY
OF THE
WORLD
OF THE
SUN.

OUR Ship at length arrived in the Harbour of *Toulon*, where the Passengers being immediately put ashore, and having thanked the Winds and Stars, for the Prosperity of our Voyage, we mutually embrac'd, and took our leave one of another. For my part, seeing in the World of the Moon, whence I came, a Song goes for Money, and that I had quite forgot the use of it, the honest Master thought himself sufficiently paid

paid for my passage, by the Honour he had of carrying on Board a Man who had dropt from Heaven: So that nothing hindred our Progress to a Friends House of mine near *Thoulouse*. I was impatient of seeing him, in hopes that I might fill him with Joy, by the Relation of my Adventures. I shall not trouble you with an account of all that happened to me upon the Road; I tired my self, and took rest; I felt hunger and thirst, and drank and eat amidst a Pack of twenty or thirty Hounds, that belonged to him. Though I was much disfigured, lean and Sun-burnt, he knew me for all that; being transported with Joy, he flew about my Neck, and having in an Extasie of Content kissed me above an hundred times, he pulled me into his House, where so soon as Tears had given way to Words: We live now, at length, cried he, and shall live, in spight of all the Accidents, wherewith Fortune hath tossed our Life. But, Good God! It was a false report then, that you were burnt in *Canada*, in that great Firework, whereof you were the Inventor? And nevertheless, two or three Persons of Credit, amongst those who brought me the sad Tidings, swore to me that they had seen and touched that Bird of Wood, wherein you were hurried away. They told me, that it was your Misfortune to go into it

at

World of the Sun.

at the very instant they put Fire to it; and that the rapid Force of the Squibs, that burnt all round it, carried you so high, that the Spectators lost sight of you: So that, as they protested, you were to that degree consumed, that the Machine falling down again, very little of your Ashes were to be found. These Ashes, Sir, then, answered I, were the Ashes of the Machine it self; for the Fire did not in the least hurt. (The Fire-works were fastened on the outside, and by consequence their Heat would not incommode me.)

Now you must know, that so soon as the Salt-peter was spent, the impetuous force of the Fire-works being no longer able to bear up the Machine, it fell to the Ground. I saw it fall;)and when I thought to have tumbled down head-long with it, I was very much surprized to find, that I mounted up towards the Moon. But I must explain to you the Cause of an effect, which you may look upon as a Miracle.

The Day when that Accident happened, I had, because of some bruises, rubbed my Body all over with Marrow: Now the Moon being then in the Wain, at which time she attracts Marrow, she suckt up so greedily that wherewith I had anointed my Flesh, especially when my box was got above the middle Region, where no Clouds in-

B 2　　　　　　ter-

terposed to weaken her Influence, that my Body followed the Attraction; and I protest, she continued to suck me up so long, that at length I arrived at that World, which here they call the Moon.

Then I told him in full all the particulars of my Voyage, and *Monsieur de Colignac*, ravished to hear things so extraordinary, adjured me to put them in Writing. I, who love Repose, declined it as long as I could, by reason of the visits that such a Publication in all probability would procure me; but being out of Countenance at the Reproach, wherewith he constantly baited me, that I made slight of his Entreaties; I resolved at length to give him that satisfaction. I put Pen to Paper then, and he being more impatiently tickled with my glory, than his own; no sooner had I made an end of a Sheet, but he hastened with it to *Thoulouse*, there to give it vent in the most ingenious Assemblies. Seeing he had the Reputation of one of the wittiest Men of the Age, my Praises, of which he was the indefatigable Herald, made me known of all Men. The Engravers, without ever having seen me, had already engraven my Picture; and the Hawkers stunned the whole City, crying about the Streets till they were hoarse again, *Who'll buy the Picture of the Author of the States and Empires of the Moon.* Amongst
those

those who read my Book, there were a great many Ignorants that were likewise medling. These that they might act the Wits of the highest flight, applauded as others did, clapt at every Word, for fear of being mistaken, and ravished with delight, cry'd, It's good! even where they understood not a tittle: But Superstition disguized into Remorse, which hath very sharp Teeth under a Fool's Coat, so knaw'd the Heart of them, that they chose rather to renounce the Reputation of a Philosopher, which, indeed, was a Habit that did not at all become them, than to answer for it at the day of Judgment.

Here, then, is the Reverse of the Medal, he's the best Man now that can retract first. The work they had so much esteemed, is no more now but a Hodge-podge of ridiculous Tales, a heap of incoherent Shreds, a Fardel of idle stories, to wheadle young Children to Bed with; and some who hardly understood the Grammar of it, condemned the Author to *Bedlam*.

This clashing of Opinions betwixt the Wise-men and Fools, encreased its Reputation. Shortly afterwards Manuscript-copies of it were sold privately; all the World, and what is out of the World also, that's to say, all from the Gentleman to the Monk, brought up the piece; nay, and the Women

men came in for a share too; every Family was divided, and the Interests of that quarrel went so far, that the whole City broke into two, the *Lunar* and *Antilunar* Factions.

Thus was the War carried on by Skirmishings, when one Morning I perceived nine or ten Beards of the long Robe enter *Colignac*'s Chamber, who presently spoke to him to this purpose. " Sir, you know that
" there is not one of us here, who is not
" your Allie, Kinsman or Friend, and that
" by consequence no Disgrace can befal you,
" but what must reflect upon us: Never-
" theless we are informed from good hands,
" that you entertain a Sorcerer in your
" House: A Sorcerer, cried *Colignac*;
" Good God! Name him to me, and I'll
" deliver him up into your Hands; but
" you must have a care it be not a Calum-
" ny. How, Sir, said one of the most ve-
" nerable, interrupting him; is there any
" Parliament more skilled in Wizards than
" ours? In a word, Dear Nephew, that
" we may hold you no longer in suspence,
" the Sorcerer whom we accuse, is the Au-
" thor *of the States and Empires of the*
" *Moon*: He cannot deny, having confessed
" what he has done, but that he is the great-
" est Magician in *Europe*. How is it possi-
" ble to mount up to the Moon, without the
" help

"help of —— ? I dare not name the Beast;
"for in short, tell me, what went he a-
"bout to do in the Moon? A pretty
"question, said another interrupting; he
"went to be present at a meeting that pos-
"sibly was kept there that day: And in-
"deed, you see he was acquainted with
"the Demon of *Socrates*. Are you surpri-
"zed then, that the Devil, as he saith,
"brought him back again into this World?
"But however it be, look ye, so many
"Moons, so many Progresses and Voyages
"through the Air, are good for nothing; I
"say nothing at all; and betwixt you and
"me, (at these words he put his Mouth to
"the others Ear) I never knew a Sorcerer
"but had Commerce with the Moon. After these good Counsels they held their
peace; and *Colignac* stood so amazed at
their Common Extravagance, that he could
not speak one word: Which a grave Cox-
comb, who had said nothing as yet, perceiving. "Look you, says he, Cousin, we know
"where the matter pinches; the Magician
"is a person whom you love, but be not
"startled, for your sake favour shall be
"shewn him; only deliver him fairly over
"to us, and in consideration of you we
"engage our Honour, to have him burnt
"without Scandal.

Colignac, at these words, though he held his sides, could not hold, but burst out into a fit of Laughter, which did not a little offend the Gentlemen his Kinsmen; insomuch that he had no power to make answer to any point of their Harangue, but by haaaa's or hoooo's; which so scandalized his worthy Relations, that they departed with shame enough to carry back with them to *Thoulouse*. When they were gone, I drew *Colignac* into his Closet, where so soon as I had shut the Door, Count, said I, to him, These long-bearded Ambassadours I don't like, they seem to me to be blazing Stars; I'm afraid the noise they have made, may be the clap of the Thunder-bolt that's ready to fall. Though their Accusation be ridiculous, and, perhaps, an effect of their Stupidity; yet I shall be no less a dead Man, though a dozen Men of Sense, who may see me roasted, should say that my Judges are Sots; all the Arguments they might use to prove my Innocence would not bring me to life again; and my Ashes would be every jot as cold in a Grave, as in the open Air: And therefore, with Submission to your better Judgment, I should joyfully consent to a Temptation which suggests to me, not to leave them any thing in this Province but my Picture: For it would make me stark-staring mad, to die for a thing which

I don't believe. *Colignac* had hardly the Patience to hear me out. However, at first he did but railly me; but when he saw that I was in earnest: Ha! s'death, cried he, before they touch a hair of your Head, I my self, my Friends, Vassals, and all that respect me, shall perish first. My House cannot be Fired without Cannon; it stands advantagiously, and is well flanked: But I'm a Fool, continued he, to caution my self against the thunder of Parchment: It's sometimes more to be feared, replied I, than the Thunder of the second Region of the Air.

From that time forward we talked of nothing but diverting our selves. One day we hunted, another we walkt and took the Air; sometimes we received Visits, and sometimes we rendred them: In a word we always changed our Recreations, before they became tiresome.

The Marquess of *Cussan*, a Man who understands the World, was commonly with us, and we with him; and to render the places of our abode the more agreeable by vicissitude, we went from *Colignac* to *Cussan*, and returned from *Cussan* to *Colignac*. The innocent Pleasures which refresh the Body, made but the least part of ours. We wanted none of those that the mind can find in Study and Conversation; and our Libraries

ries uniting like our minds, brought all the Learned into our Society. We mingled reading with Conversation; Conversation with good Cheer, that with Fishing, Hunting, or Walking; and in a word, I may say, we injoyed our selves, and whatever Nature hath produced for the Pleasure of Life, and used our Reason only to limit our Desires. In the mean time, to the prejudice of my repose, my Reputation spread it self in the Neighbouring Villages, nay and in the Towns and Cities of the Province; all Men being invited by the current Report, made a pretext of coming to see the Lord, that they might see the Sorcerer. When I went abroad, not only Women and Children, but the Men also stared at me, as if I had been the Beast. Especially, the Pastor of *Colignac*, who, whether out of Malice or Ignorance, was in secret my greatest Enemy. That Man being in appearance simple, and of a low and plain Spirit, which made him very pleasant in a kind of natural Bluntness, was in reality a very wicked Fellow: He was revengeful even to Fury; a Backbiter somewhat more than a *Norman*; and so great a Barretter, that the love of Wrangling and going to law was his predominate Passion. Having been a long time at Law with his Lord, whom he hated the more, as that he had found him firm against all his

his Attacks, he feared his Resentment; and that he might avoid it, had offered to exchange his Living: But whether he had changed his Design, or had only deferred it to be revenged on *Colignac* in my Person, during the time that he continued in his place, he strove to perswade the contrary; though the frequent Journeys he made to *Thoulouse* gave grounds to suspect it. There he told a Thousand ridiculous stories of my Enchantments; and the Suggestions of that malicious Man, concurring with the Voice of the simple and ignorant People, made my Name accursed in that place: They talked no otherwise of me than of a new *Agrippa*; and we had Information, that a Process was even commenced against me, at the suit of the Curate, who had been Tutor to his Children. This we had Notice of from several Persons, who concerned themselves in the Affairs of *Colignac* and the Marquess. And although the blockish Humor of an entire Countrey, was to us a Subject of Amazement and Laughter; nevertheless I was startled at it in private, when I more nearly considered, the troublesome Consequences that such an Error might produce. My good *Genius*, without doubt, gave me the Alarum; it enlightned my Reason with these notions, to let me see the Precipice into to which I was ready to tumble; and not thinking it enough thus tacitely to advise me, it resolved to declare more expresly in my Favors. A most troublesome Night having succeeded one of the pleasantest Days that we had spent at *Colignac*, I arose by Break of Day; and to dispell the Clouds and Cares, that still dulled and discomposed my Mind, I went into the Garden, where Verdure, Flowers and Fruits, Art and Nature, charmed the Soul through the Eyes; when at the same instant I perceived the Marquess, walking by himself, with a slow Pace, and pensive Countenance, in a large Alley, which divided the Garden into two. I was much surprised to see him, contrary to Custom, so early; that made me hasten up to him, that I might ask him the reason of it. He made me answer, that some troublesome Dreams, wherewith he had been disordered, was the Cause, that contrary to Custom, he was come so early to cure by Day, an Evil he had contracted in the Night. I confessed to him, that a like Misfortune had hindred me from sleeping, and was about to tell him the Particulars thereof; but just as I was opening my Mouth, we perceived at the Corner of a railed Walk, which crossed into ours, *Colignac* coming in great haste. So soon as he saw us at a distance, Gentlemen, cried he, take here

one

one who hath juſt eſcaped from the moſt dreadful Viſions, that are able to turn the Brains of a mortal Man. I could hardly take time to put on my Doublet, before I came down to give you an account of my Adventure; but finding neither of you in your Chambers, I haſtened to the Garden, ſuppoſing you might be there. The Truth is, the poor Gentleman was almoſt out of Breath. So ſoon as he had taken a little Breath, we entreated him to eaſe himſelf of a matter, which, though many times very ſlight, neverthelefs weighs heavy. I deſign to do ſo, replied he; but let's firſt ſit down. An Arbor of Jeſſamine offered us very pat both Seats and Shade: We entered it, and every one being placed, *Colignae* thus continued: You muſt know, that after two or three diſturbed Sleeps, I fell about Day-break into a Slumber, wherein I dream'd, that my dear Gueſt there, was in the middle, betwixt the Marqueſs and me, and that we embraced him ſtraightly, when a black Monſter, conſiſting wholly of Heads, came all of a ſudden to ſnatch him from us: Nay, I fancy'd, he was about to throw him into a great Fire, kindled hard by; for already he held him ſuſpended over the Flames: But a Virgin, like one of the Muſes, whom they call *Euterpe*, fell upon her Knees before a Lady, whom ſhe

adju-

adjured to save him (that Lady had the Presence and Marks which Painters use to give in representing of Nature.) Hardly had she heard out the Prayers of her Waiting-maid, when all amazed; Alas! cried she, he is one of our Friends. Immediately thereupon she put to her Mouth a kind of a long Pipe, like a Sackbut, and blew so long through it, under the Feet of my dear Guest, that she made him mount up to Heaven, and protected him from the Cruelties of the Monster with an hundred Heads. I fancy'd that I cried a long time after him, and adjur'd him not to be gone without me; when an infinite number of little round Angels, who called themselves the Children of the Morning, carried me to the same Countrey whither he seemed to fly, and shewed me things which I shall not relate, because I look upon them as ridiculous. We besought him, that he would tell us them however. I imagined my self, continued he, to be in the Sun, and that the Sun was a World. I had been still in the same Mistake, had not the neighing of my Horse awakned me, and convinced me that I was a Bed. When the Marquess perceived that *Colignac* had made an end: Well, then, said he, Monsieur *Dyrcona*, what was your Dream? As for mine, answered I, though it be no vulgar Dream,

yet

yet I lay no stress upon it. I am a bilious Melancholick, and that's the reason that all my life time I have dreamt of nothing but of Caves and Fire. In the prime of my Youth, I fancied in my Sleep, that I was become light, and took a flight up to the Clouds, that I might avoid the Rage of a Company of Murderers that pursued me; but that after a long and vigorous Attempt, some Wall always withstood me, though I had surmounted a great many others, at the Foot whereof, tired out with Strugling and Labour, I never failed to be stopt; or otherwise, if I imagined that I took my Flight right upwards, though I seemed for a long time to have swum in the Skies, yet I still found my self near the Earth; and contrary to all reason, though I thought my self neither weary nor heavy, yet I was still within reach of my Enemies, who stretched forth their Hands to catch me by the Foot, and pull me to them. Since I knew any thing, I never had any other Dreams but such as this, unless last night; when having, according to my Custom, flown a long while, and often escaped from my Persecutors, I thought at length, that I lost sight of them; and that in an open and clear Sky, my Body eased of all Heaviness, I pursued my Voyage into a Palace where Light and Heat are hatched. I had, without

out doubt, obſerved a great many other things; but that my Agitation to fly, brought me ſo near the Beds ſide, that I fell upon the Floor on my naked Belly, with Eyes full open. This, Gentlemen, is the ſhort and long of my Dream, which I only look upon as an Effect of thoſe two Qualities, that are predominant in my Conſtitution: for though this be a little different from thoſe which I commonly have, in that I flew up to Heaven without falling back; yet I only aſcribe that Alteration to my Blood, dilated by the Pleaſures of our Yeſterday's Diverſions, which hath diſſipated my Melancholy; and by buoying of it up, cleared it from that Weightineſs which made me tumble down again: But after all, that's a very conjectural Science. I'Faith, continued *Cuſſan*, you are in the right on't; it's a Hodge-podge of all the things we have thought on when awake, a monſtrous Chimera, a Muſter of confuſed Idea's, which the Fancy, that during Sleep, is not guided by Reaſon, preſents to us, without Order; out of which neverthleſs we think to ſqueeze the true Meaning, and draw from Dreams, as from Oracles, the Knowledge of things future; but I vow, I could never find any other Conformity betwixt them; but that Dreams, like Oracles, cannot be underſtood. However, judge of the

the worth of all the rest, by mine which is not at all extraordinary. I dreamt that I was very sad, and that I met with *Dyrcona* in all places, who called for our Assistance. But without beating my Brains any more, about the Explication of these dark Riddles, I'll tell ye their Mystical Sense in two Words; and that's, in troth, that our Dreams at *Colignac* are very bad, and that if you'll take my Advice, we'll go and have better at *Cussan*. Let's go, then, said the Count to me, since this Man is so uneasie here. We resolved to be gone the same day; and I prayed them to set out before, because I was willing, seeing (as they had agreed upon it) we were to be there a Month, to have some Books carried along with me: They condescended, and immediately after Break-fast got on Horse-back. In the mean time, I packed up some Volumes, which I imagined not to be in the Library of *Cussan*, put them upon a Mule, and about three in the Afternoon set out upon a very good Pad. However, I went but a Foot-pace, that I might attend my little Library, and at more leisure enrich my mind with the Liberalities of my sight. But listen to an Adventure, that will certainly surprise you.

I was got forwards on my Journey above four Leagues, when I found my self in a
C   Country

Country which I was certain I had seen somewhere else before: The truth is, I follicited my Memory so much to tell me, how I came to know that Landskip, that the presence of the Objects, reviving past Images, I remembred that that was exactly the place, which the Night before I had seen in a Dream. That odd rencounter would have busied my thoughts longer than it did, had I not been diverted by a strange Apparition. A Spirit, (at least I took it for one) meeting me in the middle of the way, took hold of my Horse by the Bridle. This Phantome was of a prodigious Shape, and what I could guess by the little I saw of his Eyes, had a surly and stern Look. I cannot tell, though, whether he was handsome or ugly; for a long Gown made of the Leaves of a Church plain Song-Book, covered him to the Fingers-ends, and his Face was hid under a thing like a Horn-Book, wherein was written the *in Principio*. The first Words that the Phantome uttered, were with great amazement *Satanus Diabolus, I conjure thee by the Great and Living God*, ---at these Words he stuck, but still repeating the *Great and Living God*, and with a wild and skared Look, casting about for his Pastor to blow into him the rest; when he found, that to what side soever he looked, his Pastor was not to be seen, he fell into such a dreadful shaking Fit, that by his extratraordinary chattering and diddering, one half of his Teeth dropt out, and two Thirds of the Musick-notes, under which he lurkt, flew about like Thistle Down. He came back, however, towards me, and with a Look that seemed neither soft nor surly, by which I perceived he was in doubt what course was best for him to take, whether to be rough or mild: O! well then, said he, *Satanus Diabolus, by the Blood I conjure thee in the Name of God, and of Mass-John, let me do my Business: For if thou stirrest either Hand or Foot; Devil take---thy Guts are out.* I had a lash at him with the Bridle Reins; but being almost choaked with laughter, I had little strength to do any thing: Besides that, about half a hundred Country People, came out from behind a Hedge, walking upon their Knees, and tearing their Throats with *Kyrie Eleisons*. When they were got near enough, four of the strongest of the Rout, having first plunged their Hands into a Holy Water-pot, which was purposely carried by the Priest's Man, caught hold of me by the Neck. No sooner was I arrested, but in comes *Mass John*, who devoutly pulling out his Stole, bound me fast with it; and presently after, a flock of Women and Children, who in spight of all the Resistance I could make, sowed me up in a great Sheet; wherein I was so dexteroufly fwadled, that nothing was to be seen of me but the Head. In this Equipage they carried me to *Thouloufe*; as if they had been carrying me to my Grave: By and by cried one, Had not this been done, we fhould have had a Famine, becaufe, when they met me, I was certainly going to lay a Spell upon the Corn; and then I heard another complaining, that the Scab did not begin amongft his Sheep, till of a *Sunday*, when the People were coming from *Vefpers*, I clapt him on the Shoulder. But in fpight of all my Difafters, I could hardly forbear to laugh, when I heard a young Country Girl, with a dreadful Tone, cry after her Sweet-heart, *alias* the Phantome, who had feiz'd my Horfe. (For you muft know, that the Youngfter had got on the Back of him, and fpurr'd him briskly, as if he had been his own already,) Wretch, bauled out his Duckling, What art blind then? Does n't fee that the Magician's Horfe is blacker than Coal, and that it is the Devil in Perfon carrying thee away to a meeting of Witches? Our Amorous Clown terrified at that, tumbled backwards over the Beafts Tail; fo that my Horfe was fet at Liberty. They confulted whether or not they fhould feize my Mule, and agreed in the Affirmative; but having unript the Pack, and at the opening of the firft Book hitting

upon

Descarteses Physicks, when they saw the Circles whereby that Philosopher distinguishes the Motions of the several Planets, all of them with one voice roared out, that they were the Conjuring Lines, I used to draw for raising of *Beelzebub*. He that held it in his Hands, seized with a panick fear, let it fall; and by mischance, it opened at a Page, where the Virtues of the Load-stone are explained: I say, by mischance; because, in the place I speak of, there is a Cut of that Metallick Stone, where the little Bodies, that are let loose from the whole, to fasten to the Iron, are represented like Arms. No sooner had one of the Rascals perceived it, but I heard him scream out, that that was the Toad which was found in the Manger of his Cousin *Dick's* Stable, when his Horses died. At that Word, they who seemed to be in the greatest heat, clapt their Hands into their Bosoms or Pockets. *Mass John* cried with open Mouth, that they should take special care not to touch any thing; that all these were Books of down-right Conjuring, and the Mule a *Satan*. The Rabble thus frightened, let the Mule depart in Peace. Nevertheless, I saw *Joan* the Parson's Maid drive him towards her Master's Stable, for fear he might get into the Church-yard, and there pollute the Grass of the departed.

It was full Seven of the Clock at Night, when we arrived at a Town, where for my Refreshment I was dragg'd to Goal: For the Reader would not believe me, if I said that they Buried me alive in a Hole: And neverthelefs it is true, that with one turn I furveyed the whole extent of it. In a Word, there was no Body that faw me in that place, but would have taken me for a bit of Wax-Candle, lighted under a Cupping-Glafs. At firft, when my Goaler turned me into that Cave: If you give me, faid I to him, this Stone Garment for a Doublet, it is too big; but if it be for a Tomb, it's too little. The days here are only to be reckoned by Nights; of my five Senfes, I retain only the ufe of two, Smelling and Feeling; the one, to make me fenfible of the ftink of my Prifon; and the other, to render it palpable to me. In reality, I proteft to you, I fhould think I were damned, if I knew not that no Innocent Perfon goes to Hell.

At that word Innocent, the Goaler burft out into Laughter. Nay, Faith, faid he, you are one of our right Birds then, for I never yet kept any under my Key, but fuch Gentlemen as thefe. After fome other Compliments of that Nature; the good Man took the pains to fearch me, I know not on what defign; but becaufe of the Diligence

gence he used, I conjecture it was for what I had. The pains he took in searching being all in vain, because during the Battel of *Diabolus*, I had conveyed my Gold into my Stockings; when after a most exact Anatomy, he found his hands as empty as before; both of us were within an Ace of Death, I for fear, and he for grief. S'ounds, cried he, foaming at the Mouth, at first sight I knew he was a Sorcerer, he's as poor as the Devil. Go, go, Comrade, continued he, mind the Affairs of your Conscience in time. He had no sooner said so, but that I heard the knell of a bunch of Keys, amongst which, he lookt for those of my Dungeon. His back was turned; and therefore for fear he might take his revenge for the misfortune of his Visit, I cunningly pull'd three Pistoles out of their Nest, saying to him, Master House-keeper, there's a Pistole, pray send me a bit of somewhat, for I have not eat these eleven hours past. He took it very favorably, and protested he was troubled at my Misfortune. When I perceived he was a little mollified; come, here's another, continued I, as an Acknowledgment of the Trouble, I am ashamed to give you: At once he opened his Ear, Heart, and Hand; and I added, making them up three, instead of two, that by the third I begg'd of him to let one of his Men come

and

and keep me Company, becauſe the unfortunate ought to dread Solitude.

Being raviſhed at my Prodigalities, he promiſed me all things, embraced my Legs, railed againſt the Juſtice; told me, that he well perceived I had Enemies, but that I ſhould come off with Honour: that I ſhould take good Heart; and that in the mean time, he engaged himſelf before three days were over, to have my Cuffs waſht for me. I thanked him very ſeriouſly for his Courteſie; and my dear Friend having hung about my Neck, till he had almoſt ſtrangled me, went his way, bolting and double bolting the Door.

I remained alone, and very Melancholick, lying round upon a little old Straw, reduced almoſt into Duſt. However, it was not yet ſo ſmall, but that above half a hundred Rats were ſtill a grinding of it. The Vault, Walls, and Floor, were made up of ſix Grave-Stones, that having Death over, under and about me, I might not queſtion my Enterrment. The cold Slime of Snails, and the roapy Venom of Toads, dropt upon my Face; the Fleas there had Teeth longer than their Bodies; I found my ſelf tormented with the Stone, which was not the leſs painful, becauſe it was External. In a word, I fancy that I wanted no more but a Wife, and a Pot-ſheard to make me a real *Job*.

World of the Sun.

I had, however, overcome all the Hardships of two very irksom Hours, when the noise of a Gross of Keys, with the ratling of the Bolts of my Door, diverted me from minding my Pains. After the jingling noise, by a little Lamp-light, I perceived a sturdy Clown. He unloaded an earthen Dish between my Legs. And there, there, said he, be not disturbed, there's a good Cabage Soop for ye; and were it—— but indeed it is my Mistress's own Soop; and faith and troth, as the saying is, there is not one drop of the Fat taken off on't. Having said so, he dives his four Fingers and Thumb to the very bottom of the Dish, to envite me to do the like. I followed my Copy, for fear of discouraging him; and he with a joyful glance of an Eye, S' diggers, cried he, you are an honest Brother. They zay you've got Illwillers: S'lid they are Traytors; yes Dad, they are very Traytors: Well, wou'd they'd come here and see. Ay, ay, it is so; he goes first that leads the Dance. This blunt Simplicity brought a fit of Laughter two or three times up to my very Throat. However I was so happy as to check it: I perceived, that Fortune, by means of this Rogue, seemed to offer me an occasion of Liberty; and therefore it extreamly concerned me to gain his Favor; for otherwise to escape, it was impossible. The Architector that built my my Prison, having made my Entries into it, did not bethink himself of making one Outlet. These Considerations were the Cause, that to sound him, I spake to him to this purpose, My good Friend, thou art a poor man, is n't that true? Alas! Sir, answered the Clown, had you been with the cunning Man, you could not have hit righter. Here then, said I, take that Pistole.

I found his hand to shake so, when I put the Pistole into it, that scarcely could he shut it. That begining seemed to me to be a little ominous: However, I quickly perceived by the heartiness of his Thanks, that he only trembled for Joy; and that made me go on: But wert thou a man, that would be concerned in the accomplishment of a Vow which I have made, besides the Salvation of thy Soul, thou might'st be as sure of twenty Pistoles, as thou art of thine own Hat: For thou must know, that it is not as yet a full quarter of an Hour, in a word, a moment before thou camest, that an Angel appeared to me, and promised to make the Justice of my Cause appear, provided I went to morrow to our Lady's Church of this Town, and had a Mass said at the high Altar there. I pretended to excuse my self upon the account of my close Imprisonment; but the Angel made Answer, That a man should come, sent from the Goaler, to keep me

com-

company, whom I should but command in his name to carry me to Church, and bring me back again to Prison; that I should enjoyn him Secresie, and to obey without gainsaying, upon pain of dying within the Year; and if he questioned the Truth of what I said, I should give him this token that he had been touched for the Evil. Now the Reader must know, that I had seen through a hole of his Shirt, a piece of the King's Gold, which suggested to me the whole Series of this Apparition. Yea, verily then, said he, good Sir, I shall even do what the Angel has commanded me; but it must be at nine of the Clock in the Morning, because at that time our Master will be at *Thoulouse*, about the making up of a Match betwixt his Son, and the *Master of the last Work*'s Daughter. D'ye mind me, Sir, the Hangman has a Name, as well as a Crab-louse: They talk as if she should have from her Father, as many Crowns in Portion, as might make up a King's Ransom. In short, she is Fair and Rich; but such Wind-falls seldom fall in the way of a poor Young-Man. Alass! Good Sir, --- wou'd have you know——I failed not here to interrupt him; for I foresaw by the beginning of this Digression, that I should be baited by a long tale of a Tub. Our Plot being very cautiously laid betwixt us, the

Clown

Clown took leave of me; and failed not next Morning to come at the prefixt hour, and untomb me. I left my Cloaths in the Prison, and dressed my self in Rags; for least I might be known, we had so ordered it the Night before. So soon as we were abroad in the open Air, I forgot not to tell him down his twenty Pistoles. He looked and stared upon them very wistfully. They are good Gold and of full weight, upon my Word, said I to him. Ha, Sir, replied he, that's not the thing I mind; but I'm thinking that *great Ralph*'s House is to be sold, with a Close and Vineyard. I can have it for two hundred Francks, it will require eight days time to make up the Bargain; and I would beg of you, good Sir, if it be your Will and Pleasure, so to order the matter, that till *great Ralph* have told and received your Pistoles, and lockt them safe up in his Chest, they may not turn into Shells. I could not but laugh at the simplicity of the Knave. In the mean time, we jogg'd on towards the Church, where at length we arrived. Shortly after high Mass began; but so soon as I saw my Keeper rise in his turn to go to the offering, I skipped at three leaps out of the Church, and at as many more whipt into a little Bye-Street or Alley. I had a great many thoughts in my head at that instant; but that which

I

I followed was to get to *Thoulouse*, which was but half a League distant from the place, with Design to take Post there. I got to the Suburbs in very good time; but I was so ashamed to be stared at by all the People that saw me, that I was quite out of Countenance: That which made them stare, was my Dress; for I being but a Novice in the begging Trade, had marshall'd my Clouts about me so odly, that with a Gate that suted not at all with my Habit, I seemed to be one in Disguise, rather than a Beggar; besides that, I made great haste, looked down, and asked nothing. At length, considering that this general Observation of the People threatned me with some dangerous Issue, I overcame my Bashfulness. So soon as I perceived any one that lookt at me, I stretched him out my Hand: Nay, I even importuned the Charity of those that did not in the least mind me. But reflect a little, and wonder how many times by using too great Circumspection, about the Designs wherein Fortune will have some share, we spoil them by provoking that haughty Goddess. I make this Observation, upon Occasion of the Accident that befel me; for perceiving a Man in the Dress of an ordinary Citizen, with his Back towards me, Sir, said I, pulling him by the Cloak, if there be any Bowels of pity——

I had not brought forth the word that was to come next, when the Man turned about his Head. Good God! What was he? Nay, Good God! What was I? That very Man was my Goaler; we stood both amazed with Admiration to see one another in the place we did. His Eyes were wholly fixed on me, and I had nothing in view but him. In fine, a common Interest, though very different, recovered us out of the Extasie wherein both of us were plunged. Ha! Wretch that I am, cried the Goaler, must I then be catcht? That Word of a double meaning put into my mind the Stratagem you shall hear. Stop Thief, Gentlemen, stop Thief, cried I, as loud as I could bauls This Rogue hath stole the Jewels of the Countess of *Monseanx*; I have been a year in search of him. Gentlemen, continued I, all in a heat, a hundred Pistoles for him that shall take him. No sooner had I let fly these Words, but a Troop of the Rabble fell upon the poor amaz'd Wretch. The Surprize that my impudence had cast him into, being heightened by the Imagination he had, that without a Body, like unto that of the Saints in Glory, which might pass entire through the Walls of my Dungeon, I could not have made my escape, so transported him, that he was for a long time besides himself. He came to himself again, however,

ever, at length; and the first Words he used to undeceive the Mobile, were, That they should have a care they did not commit a mistake; that he was a man of Honour and Reputation. Without doubt he was about to discover the whole Mystery: But a dozen of Coster-mongers, Lackeys and Chair-men, being desirous to serve me for my Money, stopt his Mouth with Fisty-cuffs; and in as much as they fancied, that their *Reward* should be proportioned to the degree of Insolence, wherewith they insulted over the Weakness of the poor stunn'd Man, every one came running in to have a touch at him, either with Hand or Foot. Here's your Man of Honour, cried the Riff-raff, and yet he could not forbear to say, so soon as he knew the Gentleman, that he was catcht. The Cream of the Jest was, that my Goaler being in his Holy-days Cloaths, was ashamed to confess himself to be the Hang-man's Church-warden; nay, he was afraid that by discovering himself to be what he was, he might but encrease the number of his Blows. For my own part, whilst the scuffle was at the length I took my flight. I trusted my safety to my Legs, which would have soon set me at Liberty: But as the Devil would have it, the People beginning of new all to stare at me, I found my self in as bad a pickle again as at first. If the Spectacle

of

of an hundred Rags, which like a Brawl of little Beggars danced about me, did excite the Curiosity of any gaping Lout to stare at me; I was afraid that he might read in my Fore-head, that I was one that had broken Prison. If any one passing by me, put his Hand out under his Cloak, I fancy'd him to be a Serjeant, who stretched out his Arm to lay hold on me. If I observed another scampering along the Streets, without casting an Eye on me, I perswaded my self that he feigned not to see me, to the end he might snap me behind. If I perceived a Shop-keeper enter his Shop, now, said I, he is gone to fetch out his Constables Staff. If I came into a place where there was any extraordinary Concourse of People, so many Men, thought I, could not be got together there without some Design. If another place was empty, here they lye in wait for me. Did I meet with a stop, now, thought I, they have barocado'd the Streets to shut me in. In a word, Fear perverting my Reason, every Man seem'd to me to be an Officer, every Word, Stop, and every noise, the insupportable Rattling of the Bolts of my last Prison. Being thus beset with panick Fear, I resolved to play the Beggar again, that so I might pass the rest of the City, till I got to the Post-house: But fearing lest my Voice might betray me,

World of the Sun.

I thought best to Counterfeit the Dumb-man. I advanc'd then towards those whom I pereived to Eye me. I pointed with my Finger under my Chin, then over my Mouth, and gaping made an unarticulate Cry, to give them to understand by this Action, that a poor Dumb-man begg'd their Charity. Sometimes I had a compassionate Shrug of the Shoulders for an Alms; by and by I felt a small Bribe slipt into my Fist; and anon again I could hear the good Women mutter, that perhaps I might have been in that manner Mortifyed for the Faith in *Turkey*. In short, I learnt that the begging Trade is a great Book, that Instructs us in the manners of People at a cheaper rate, than all the long Voyages of *Columbus* and *Magellan* can

That Stratagem however, could not as yet prevail over the Head-strongness of my Destiny, nor overcome the ill nature thereof: But what other Invention could I betake my self unto? For to cross so great a City as *Thoulouse* is, where my Garb had made me known even to the Herring-wives, having more shaggy Rags dandling about me, than the errantest Tatterdemallion in the World, was it not very likely that I should immediately be taken notice of and known? And that the only charm against that danger was to personate the Beggar, whose

whose part is Acted under all Shapes? And then granting this Trick had not been projected with all necessary Circumspection, I fancy still that amongst so many fatal Junctures, it was a sign of a very good Judgment, not to run stark mad.

I was setting forward on my Journey then, when all of a sudden I found my self obliged to turn back again; For my venerable Goaler, and about of a dozen of Officers of his Acquaintance, who had rescued him out of the Hands of the Rabble, setting out upon the Hunt, and scowring all the Town to find me, fell unluckily in my way. So soon as with Eagles Eyes they perceived me, you may imagine, that they run, and I run, with all the Speed we could. I was so nimbly pursued, that sometimes my Liberty fell upon its Neck, the Breath of the Tyrants who sought to oppress it: But it seemed that the Air, which they pushed forwards running after me, drove me on before them. At length kind Heavens or Fear rather carried me four or five Lanes on Head of them. Then it was that my Hunters lost the scent, and I the view and shameful Noise of that troublesome Chace. Certainly he, that hath not escaped such like Agonies, I speak by Experience, can hardly measure the Joy wherewith I was transported, when I found my self out of their Clutches.

Clutches. However, seeing my Safety required all my skill, I resolved avaritiously to Husband the time which they spent in dogging of me. I besmeared my Face, rubbed my Hair with Dust, stript off my Doublet, let fall my Breeches, threw my Hat in a Cellar; and then having spread my Handkerchief upon the Street, with four little Stones on the Corners, as they do who are infected with the Plague, I laid my self down upon my Belly over against it, and with a lamentable Tone fell a Groaning most languishingly. Hardly was I placed in this manner, when I heard the cry of the wheezing Rabble, long before I heard the sound of their Feet; but I had still Judgement enough to keep my self in the same posture, in hopes that I might not be known; and I was not mistaken, for all taking me for one Infected, they passed by me in great haste, stopping their Noses, and most of them throwing a Double upon my Handkerchief.

The storm being thus over, I slipt into an Ally, put on my Cloths again, and once more trusted my self to Fortune; But I had run so long that she was weary of following me. No body could think otherwise, for having scuddled over so many publick Places and Quarters of the Town, tript along and turned so many Streets, that lofty

D 2   God-

Goddess, unaccustomed to march so fast, to put a stop to my Carriere, suffered me blindly to fall into the Hands of the Officers that pursued me. At our meeting they thundred out so loud an Hue and Cry, that I was quite stunn'd with the Noise. They thought they wanted Arms enough to hold me, and therefore employed their Teeth, not believing they had me sure enough; one dragged me by the Hair, another by the Collar, whilst the less passionate rifled me; and had better luck than my Goaler at the first search, for they found the rest of my Gold.

Whilst these Charitable Physicians were taken up in curing the Dropsie of my Purse, a great Hubbub arose; all the place resounded with these Words, *kill, kill*; and at the same time I saw drawn Swords. The Gentlemen who dragg'd me along, cried that they were the Officers of the City Magistrates, who had a mind to take their Prisoner from them. But take heed, said they to me, tugging me along with greater Force, that you do not fall into their Hands, for if so, you'll be condemned within four and twenty Hours, and then the King cannot save you. At length, however, they themselves being afraid of the Rout, that began to come up with them, left me so universally, that I remained all alone in the middle

dle of the Street, whilst the Aggressors in the mean time, butchered all they met with. I leave it to you to judge, whether or not I betook my self to my Heels, having cause to be equally afraid of both. In a trice I was got at a distance from the Hurry; but just as I was asking the way to the Post-House, a torrent of People that fled from the Scuffle broke into my Street; being unable to resist the Croud, I followed it; and being vexed to run so long, I gained at length a little dark Gate, into which I threw my self pell-mell with those that fled. We shut it upon our selves; and then when we had all taken Breath: Comrades, said one of the Gang, if you'll take my advice, let us pass the two Wickets, and make for the Court. These dreadful Words struck me with so surprising a Grief, that I thought to have fallen dead upon the place. Alas! *I* perceived immediately, but too late, that instead of saving my self, as *I* thought in a Sanctuary, *I* had cast my self into Prison; so impossible it is to avoid the Influence of ones Watchful Stars. *I* lookt upon that Man more attentively, and knew him to be one of the Officers, who had so long given me the Chace: *I* fell into a cold Sweat, and lookt Pale as if *I* had been ready to faint away. They who saw me in so weak a Condition, being moved with Compassion,

passion, call'd for Water; every one drew nigh to assist me; and by mischance that accursed Officer was one of the first: He had no sooner beheld me, but that he knew me. He made a Sign to his Companions, and at the same time *I* was saluted with an *I Arrest you Prisoner in Name of the King*. They needed not go far to enrol my Name.

I remained in the Cage till Night, where every Turn Key one after another, by a exact Dissection of the Parts of my Fac drew my Picture upon the Cloth of h Memory.

At seven a Clock at Night the jingling a Bunch of Keys gave the signal of Retrea They asked me if *I* would be carried t a Chamber of a Pistole; *I* answered with nod of the Head. Money then, repli the Guide. *I* knew very well *I* was in place where *I* must pocket a great man such Snubs: And therefore *I* prayed him, case he could not be so Courteous, as to gi me Credit till next Morning, that he woul tell the Goaler from me, he should restor me the Money that had been taken fro me. Ho, ho! I'faith, answered the Villai our Master is a Man of Heart; he gives n thing back. Doe ye think then that for th sake of your pretty Nose.---along, along, t the Dungeon. Having said so, he shew

World of the Sun.

me the way by a lusty Thump with his Bunch of Keys; the weight whereof made me tumble and slide from the top to the bottom of a dark Ascent, till *I* knocked against a Door that stopt me: Nor, indeed, had *I* known it to be a Door, but for the rap *I* gave against it: For *I* had not now my Eyes, they remained at the Stairs-Head under the Figure of a Candle, which my Hang-man Guide held in his Hand fourscore steps above me. At length that Tyger of a Man being come down *Pian Piano*, unlocked thirty great Locks, pull'd out as many Bars; and the Wicket being only half opened, with a joult of his Knee he ingulfed me in that Pit, whereof *I* had not time to observe the Horrour, so suddenly he pulled the Door after him. *I* stood in mire up to the Knees. If *I* had a mind to get to the side, *I* fell in up to the middle. The terrible clucking of the Toads that crawled in the Vessel, made me wish my self Deaf; *I* felt Asks creeping by my Thighs, Serpents twisting about my Neck; and one *I* espied by the somber light of his sparkling Eyes, from a Mouth black with Venom, darting a forked Tongue, whose brisk Agitation made it look like a Thunder-bolt, set on Fire by its Eyes.

I cannot express the rest; it passes all belief; and besides, *I* dare not reflect upon the same; so afraid I am, that the Assurance I think my self in, of being freed from my Prison, should be no more but a Dream, out of which I am ready to awake. The Gnomon had marked Ten of the Clock upon the Dial of the great Tower, before any Body came to knock at my Tomb: But about that time, when bitter Grief and Sorrow began already to press my Heart, and discompose that just Harmony wherein consists Life, I heard a voice that bid me take hold of the Pole that was presented unto me. Having a long time felt about in the dark to find it, at length I met with one end thereof; with extraordinary motion I took hold on't, and my Goaler pulling the other end towards him; angled me out of the middle of that Mire. I began to suspect that the Countenance of my Affairs was changed, for he shew'd me great Civility, spoke to me bare-headed, and told me that five or six Persons of Quality waited in the Court to see me. Amongst the rest, not so much as that wild Beast who shut me up in the Den, which I have described to you, but had the Impudence to accost me, with one Knee on the Ground, having kissed my Hand, he beat off a great many Snails that stuck to my Hair with one of his Paws, and with the other a great cluster of Leeches, wherewith my Face was Vizor-masked.

Having

Having performed this rare piece of Civility; at least, Good Sir, said he to me, you'll think on the Care and Pains that great *Nicolas* has taken about you: S'death, d'ye mind me, when it was done for the King, it is not for you to upbraid him for it, I trow. Being madded at the Impudence of the Rascal, I made him a Sign that I should think on't. Through a Thousand dreadful turnings, at length I came into the Light, and afterwards into the Court, where as soon as I entred it, two Men caught hold on me, whom at first I could not know, by reason they fastened about my Neck at the same time, and joined their Faces close to mine. It was a pretty while before I could guess who they were; but the Transports of their Friendship intermitting a little, I knew my dear *Colignac*, and the brave Marquess. *Colignac* had his Arm in a Scarf, and *Cussan* was the first that came out of his Extasie. Alas! said he, we had never suspected such a disaster, had it not been for your Horse and Mule, who that Night came to my Gate: Their Girths, Cruppers, and all were broken, and that made us presage some Misfortune was befallen you. We presently got on Horse-back, and had not rid two or three Leagues towards *Colignac*, when all the Country alarm'd at that Accident, told us the particular Circumstances there-

thereof. We presently gallop'd to the Town, where you were in Prison; but being there informed of your escape, upon the rumor that went, that you had taken your course towards *Thoulouse*, with what men we had, we posted thither in all haste. The first man we asked news of you; told us that you were retaken, at the same time we spurred our Horses towards this Prison; but others assured us, that you had vanished out of the Hands of the Serjeants: And as we still went on, the Towns people were telling one another, how you were become invisible. At length having made further and further inquiry, we came to know that after you had been taken, lost, and retaken, I know not how many times, you were carried to Prison, in the great Tower. We way-laid your Officers, and by good Fortune, through more apparent than real, met, attacked, beat and put them to Flight; but we could not learn, even of the Wounded whom we took, what was become of you; until this Morning word was brought us, that you your self had blindly secured your self in Prison. *Colignac* is wounded in several places, but very slightly. After all, we have just now taken order, that you be lodged in the fairest Chamber that's here: Seeing you love an open Air, we have caused to be furnished

a

a little Appartment for you alone in the top of the great Tower, the Terrass whereof will serve you for a Balcony; your eyes, at least will be at Liberty, in spight of the Body they are fastened to. Ha! my Dear *Dyrcona*, cryed the Count speaking next; we were unfortunate we did not take you along with us, when we parted from *Colignac*: My Heart by an unaccountable Sadness, that I could give no reason for, presaged some terrible Disaster; but it matters not, I have Friends, thou art Innocent, and let the worst come to the worst, I know what it is to dye Gloriously. One thing only puts me in despair. The Villain on whom I resolved to try the first stroke of my Revenge, (you well conceive I speak of my Curate) is now out of condition of feeling it; the wretch is dead, and I'll tell you the particulars of his death: He was running with his Man to drive your Horse into his Stable, when the Nagg with a fidelity heightened perhaps, by the secret notices of his Instinct, falling into a sudden Fury, began to winse and kick; but with so much rage and success, that with three kicks of his heels he made Vacant the Benefice of that Bufflehead. Without doubt you cannot conceive the Reasons of that Fools hatred, but I'll discover them to you: Know then that I may trace the matter a little backward,

that

that that Godly man, a *Norman* by Nation, and a litigious Knave by Trade, who for the Money of Pilgrims officiated in a forsaken Chappel, commenced an Action of Devolution against the Curate of *Colignac*; and maugre all my endeavours to maintain the Possessor in his right, so wheadled the Judges, that at length in spight of us, he was made our Pastor.

At the end of the first year, he went to Law with me also, pretending that I should pay him Tythes: It was to no purpose to tell him, that time out of mind my Lands were free; he went on still with his Suit, which he lost. But during the Process, he started so many Cases, that Twenty other Suits have sprung from them, which now are at a stand; thanks to the good Horse whose Foot was harder than *Mass-Johns* Head. This is all that I can conjecture of the Vertigo of our Pastor. But it's wonderful with how much fore-sight he managed his Rage: I am lately assured, that having got into his Head the accursed design of your Imprisonment, he had secretly exchanged his Living of *Colignac*, for another Living in his own Country, whither he intended to retreat so soon as you should be taken: Nay his own Man hath said, that seeing your Horse near his Stable, he had heard him mutter, That the Beast would carry ry him into a place, where they could not reach him.

After this Discourse, *Colignac* admonished me to mistrust the Offers and Visit, that perhaps might be rendred me by a very powerful Person, whom he named; that it was by his Credit, that *Mass-John* had gained the Cause of Devolution; and that that Person of Quality, had sollicited the Affair for him in recompence of the Services, which that good Priest had rendred his Son, when he bore a small Office in the Colledge. Now, continued *Colignac*, seeing it is very hard to be at Law without Rancor, and without a tincture of Enmity, that remains indelible in the mind; though we have been made Friends, he hath ever since sought occasions secretly to cross me: But it matters not, I have more Relations of the long Robe than he has, and a great many Friends, or if it come to the worst, we can procure the King to interpose his Authority in the Affair.

When *Colignac* had made an end, they both endeavoured to Comfort me; but it was by such tender Testimonies of Sorrow, that my own Grief was thereby encreased.

In the mean while my Goaler came back, and told us that the Chamber was ready. Come let's go see it, answered *Cussan*; and with that he went first, and we followed him:

him: I found it in very good Order: I want nothing, said I to them, unless it be a few Books. *Colignac* promised to send me next day, as many as I should give him a Catalogue of. When we had well considered and found by the height of my Tower, the largeness of the Ditches that environed it, and by all the Circumstances of my Apartment, that to escape was an enterprise above humane reach; my Friends looking on one another, and then casting their Eyes on me, fell a weeping. But as if all of a sudden our Grief had softened the Anger of Heaven, an unexpected Joy took possession of my Soul; Joy brought Hope, and Hope secret Illuminations, wherewith my Reason was so dazled, that with an unvoluntary Transport, which seemed ridiculous to my self: Go, said I to them, go expect me at *Colignac*; I shall be there within these three days; and send me all the Mathematical Instruments wherewith I usually work: In short, you'l find in a large Box, a great many peices of Christal cut into several Figures, be sure not to forget them; however, it will be sooner done, if I set down what things *I* need in a Memorandum.

They took the Note *I* gave them, being unable to dive into my design; and then departed.

World of the Sun.

From the time they were gone, *I* did nothing but ruminate upon the Execution of the things *I* had premeditated, and *I* was thinking on them next day, when all that *I* had set down in my Catalogue was brought me from them: One of *Colignac*'s *Valets de Chamber* told me, that his Master had not been seen since the day before; and that they could not tell, what was become of him. I was not at all troubled at that Accident, because it presently came into my mind that possibly he might be gone to Court to sollicite my Liberty: And therefore without being surprised at it, I put hand to work; for the space of eight days, I hewed, plained and glewed, at length I framed the Machine, that I am about to describe to you.

It was a large very light Box, that shut tight and close; of about six Foot high, and three Foot Square. This Box had a hole in it below; and over the Cover, which had likewise a hole in it, I placed a Vessel of Christal, bored through in the same manner, made in a Globular Figure, but very large, the Orifice whereof joyned exactly to and was enchaced, in the hole I had made in the head.

The Vessel was purposely made with many Angles, and in form of an Icosaedron, to the end that every Facet being convex and concave, my Boul might produce the effect of a Burning-Glass. The

The Goaler, and his Turn-keys never came up to my Chamber, but they found me employed in this work; but they were not at all surprised at it, because of the many Mechanick Knacks which they met with in my Chamber, whereof I told them I was the Inventor: Amongst others there was a Wind-Clock, an Artificial Eye, wherewith one might see by night; and a Sphere wherein all the Stars followed the regular motion that they have in the Heavens: By these things they were perswaded, that the Machine I was a making, was a Curiosity of the like Nature; and besides the Money wherewith *Colignac* greased their fists, made them go fair and soft. Now it was about nine in the Morning; my Keeper was gone down, and the Skie was hazy, when I placed this Machine on the top of my Tower, that's to say, on the openest place of my Terrass walk: It shut so close, that a grain of Air could not enter it, except by the two openings; and I had placed a little very light Board within for my self to sit upon.

Things being ordered in this manner, I shut my self in, and waited there almost an hour, expecting what it might please Fortune to do with me.

When the Sun breaking out from under the Clouds, began to shine upon my Machine

chine, that transparent Icosaedron, which through its Facets received the Treasures of the Sun, diffused by it's Orifice the light of them into my Cell; and seeing that splendor grew fainter, because of the Beams that could not reach me, without many Refractions, that tempered vigour of light converted my Case into a little Purple Firmament, enameled with Gold.

With extasie I admired the Beauty of such a mixture of Colours; when all of a sudden *I* found my Bowels to move in the same manner, as one finds them that is tossed in a swing.

I was about to open my Wicket, to know the cause of that emotion; but as *I* was stretching out my Hand, through the hole of the Floor of my Box, *I* perceived my Tower already very low beneath me; and my little Castle in the Air, pushing my Feet upwards, in a trice shew'd me *Thoulouse* sinking into the Earth. That Prodigy surprised me; not at all by reason of so sudden a soaring, but because of that dreadful transport of Humane Reason, at the Success of a design, which even frightned me in the Project. The rest did not at all Startle me; for *I* foresaw very well, that the Vacuity that would happen in the Icosaedron, by reason of the Sun-Beams, united by the concave Glasses, would, to fill up the space,

E   attract

attract a great abundance of Air, whereby my Box would be carried up; and that proportionably as *I* mounted, the rushing wind that should force it through the Hole, could not rise to the roof, but that furiously penetrating the Machine, it must needs force it up on high. Though my design was very cautiously projected, yet *I* was mistaken in one circumstance; because *I* was not confident enough of my Glasses. *I* had prepared round my Box a little Sail, easie to be turned, with a Line that passed through the Orifice of the Vessel; and which *I* held by the end; *I* had fancied to my self, that when *I* should be in the Air, *I* might thus make use of as much wind, as might serve to convey me to *Colignac:* But in the twinkling of an Eye, the Sun which beat perpendicularly, and obliquely upon the Burning-Glasses of the Icosaedron, hoisted me up so high, that *I* lost sight of *Thoulouse.* That made me let go my sheet, and soon after *I* perceived through one of the Glasses, which *I* had put in the four sides of the Machine, my Sail flying in the Air, and tossed to and fro by a Whirl-wind that had got within it.

I remember, that in less than an hour *I* was got above the Middle Region; and *I* soon perceived it, because *I* saw it hail and rain below me: It may be asked, perhaps, whence

whence then came that wind (without which my Box could not mount) in a story in the Sky exempt from Meteors; but provided I may have a hearing, I'll answer that Objection. I have told you, that the Sun which beat vigorously upon my Concave Glasses, uniting his Rayes in the middle of the Vessel, by his heat drove out the Air it was full of through the upper Conduit; and that so the Vessel being void, Nature, which abhors Vacuity, made it suck in, by the opening below, other Air to fill it again: If it lost much, it regained as much; and so one is not to wonder, that in a Region above the middle where the winds are, I continued to mount up; because the Æther became wind, by the furious Rapidity wherewith it forced in to hinder a Vacuity, and by consequence ought incessantly push up my Machine.

I felt little or no Hunger, except when I passed that Middle Region of the Air; for in reality the coldness of the Climate, made me see it at a distance: I say at a distance, because a Bottle of Spirits which I carried always about me, whereof I now and then took a dram, kept it from approaching me.

During the rest of my Voyage, I felt not the least touch of it; on the contrary the more I advanced towards that enflam'd World, the stronger I found my self. I felt my Face to be a little hotter and more gay than ordinary; my Hands appeared to be of an agreable Vermilion Colour, and I know not what Gladness mingled with my Blood, which put me beyond my self.

I remember, that reflecting once on this Adventure, I reasoned thus with my self. Hunger without doubt cannot reach me, because that pain being but an Instinct of Nature, which prompts Animals to repair by Nourishment, what they lose of their Substance: At present when she finds, that the Sun by his pure, continual and neighbouring Irradiation, stocks me with more natural Heat than I lose; she gives me no more that Desire, which would be useless. Nevertheless I objected against those Reasons, that seeing the Temperament which maketh Life, consisted not only in natural Heat, but also in radical Moisture, on which that heat is to feed, as the Flame in the Oyl of a Lamp: The sole Rays of that vital Fire, could not make Life; unless they encountered some unctuous Matter that should fix them. But I presently overcame that difficulty, when I had observed, that in our Bodies the radical Moisture and natural Heat are but one, and the self same thing; for that which is called Moisture, whether in Animals or in the Sun, that great Soul of the World, is but a flux of Sparkles; more
conti-

continuous because of their Mobility; and that which we name Heat, a Concourse of Atomes of Fire, which appear looser because of their interruption; but though the radical Moisture and natural Heat were two distinct things, yet it is certain, that the Moisture would not be necessary for living so near the Sun; for seeing that Humidity in living Creatures, serves only to detain the heat, which would exhale too fast, and could not be restored so soon; I was in no danger of wanting it, in a Region whereof these little Bodies of Flame which constitute Life, more of it was united to my Being, then separated from it.

There's another thing that may be wondered at; and that is, why the approaches of that burning Globe consumed me not, for I was already got almost within the full Activity of its sphere; but I have a reason at hand for that. To speak properly, it is not the Fire it self that burns, but a grosser matter, tossed to and fro by the dartings out of it's moveable Nature; and that Powder of little Sparks, which I call Fire, moveable of it self, owes, possibly, all it's Action to the Roundness of it's Atomes; for they tickle, warm, or burn, according to the Figure of the Bodies, which they draw along with them. So Straw sends not forth so burning a Flame as Wood does; Wood burns with less violence than Iron; and the Reason of this is, because the Fire of Iron, of Wood, and of Straw, though in it self it be the same Fire, yet it acts variously, according to the diversity of the Bodys that it moves: And therefore in Straw, the Fire (that spiritual Dust, as it were,) being entangled only with a soft Body, is less corrosive: In Wood, whose substance is more compact, it enters more hardly; and in Iron, the Mass whereof is almost wholely solid, and knit together by Angular Particles, it penetrates and consumes in a trice whatsoever it touches. These Observations being also familiar, no Body will wonder, that I approached the Sun without being burnt; seeing that which burns is not the Fire, but the matter to which it is joyned, and that the Fire of the Sun, cannot be mingled with any matter. Don't we even find by experience, that Joy which is a Fire, because it only moves an Aërial Blood, whose subtile Particles beat gently against the Membranes of our Flesh, tickles and produces I know not what blind Pleasure; and that that Pleasure, or rather that first Progress of Pain, advances not so far as to threaten the Animal with Death; but only to make him sensible, that Desire causes a Motion in our Spirits, which we call Joy. Not but that a Feaver, though it have quite

con-

contrary Symptoms, is a Fire as well as Joy; but it is a Fire, wrapped up in a Body of Horned Particles, such as the *Atra bilis* or Melancholy, which darting out his hooked Points, in all parts where its movable Nature carries it, pierces, cuts, excoriates, and by that violent Agitation, produces that which is called the Heat of a Feaver; but this Concatenation of Proofs is useless; the most vulgar Experiments are sufficient to convince the obstinate. I have no time to lose, I must look to my self: I am like *Phaeton* in the middle of a Career, where I cannot turn back again; and wherein if I make but one false step, all Nature is not able to save me.

I perceived most distinctly, what heretofore I suspected, when I went up to the Moon; that, indeed, it is the Earth that moves from East to West about the Sun, and not the Sun about it: For *I* saw next to *France*, the Foot of the Boot of *Italy*, then the *Mediterranean-Sea*, then *Greece*, then the *Bosphorus*, the *Euxin-Sea*, *Persia*, the *Indies*, *China*, and at length *Japan*, pass successively over against the hole of my Lodging; and some Hours after my Elevation, all the South-Sea having turned, gave way to the Continent of *America*. I clearly distinguished all these Revolutions; nay, and I remember that a long time after, *I* saw *Europe* mount once more again upon the Scene; but *I* could not now obferve the feparate States thereof, becaufe *I* was exalted too high. *I* left by the way, fometimes on the right, and fometimes on the left, feveral Earths like ours, where if *I* touched the leaft upon the Spheres of their Activity, I found my felf incline fide-ways: However, the rapid force of my foaring Flight, furmounted the Power of thefe Attractions.

I coafted by the Moon, which at that time was got betwixt the Sun and the Earth, and *I* left *Venus* on my right-hand. But now *I* am fpeaking of that Star, the old Aftronomy hath fo long preached, that the Planets are Stars which turn round the Earth, that the Modern dares not make a doubt of it: And neverthelefs, *I* obferved, that during the whole time, that *Venus* appeared on this fide the Sun, about which fhe turns, *I* faw her always Crefcent; but finifhing her courfe, *I* obferved that as fhe paft gradually behind him, the Horns drew nearer, and her Black Belly was guilt again. Now this viciffitude of Light and Darknefs, makes it very evident, That the Planets are like the Moon and the Earth Globes without Light, capable only to reflect that which they borrow.

Th

The Truth is, as *I* still mounted, *I* made the same Observation of *Mercury*. *I* observed besides, that all these Worlds have other little Worlds also which move about them. Musing afterwards on the Causes of the Construction of this great Universe, I imagined that at the clearing of the *Chaos*, after that God had created the Matter, Bodies of the like Nature joined together, by that Principle of unaccountable Love, by which we see by Experience that every thing covets its like; Particles formed after a certain Fashion assembled together, and that made Air: Others again, to which the shape gave a circular Motion, by clinging fast together, composed the Globes that are called Stars, which not only, because of that Inclination of whirling about upon their Poles, to which their Figure determines them, ought to truss together into a Round as we see them; but ought in the same manner, as they evaporated from the Mass, and kept a Parallel March in their flight, make the lesser Orbs, which they met in the Sphere of their Activity, to turn: And therefore *Mercury*, *Venus*, the Earth, *Mars*, *Jupiter*, and *Saturn*, have been constrained to whirlegig it, and move both at once about the Sun. Not but that one may imagine, that heretofore all those other Globes have been Suns, since the Earth still retains
in

in spight of its present Extinction, heat enough to make the Moon turn about it, by the circular motion of the Bodies, which are detach'd from its Mass, and that *Jupiter* retains enough to make four to turn: But these Suns by the length of time, have sustained so considerable a loss of Light and Fire, by the continual Emission of the little Bodies, which make Heat and Light; that they remain no more but a cold, dark, and almost unactive *Caput Mortuum*. Nay, we discover that these spots which are in the Sun, and were not perceived by the Ancients, dayly increase: Now, who can tell but that it is a Crust formed in its Superfice, it's Mass that extinguishes proportionably as the Light leaves it; and if it become not, when all these moveable Bodies have abandoned it, an obscure Body like the Earth? There are very distant Ages, beyond which there appears no Vestige of Man-kind: perhaps heretofore, the Earth was a Sun peopled with Animals, proportioned to the Climate that produces them; and perhaps these Animals, were the *Demons* of whom Antiquity relates so many Instances. Why not? Is it not possible, that these Animals after the Extinction of the Earth, have still lived there for some time, and that the Alteration of their Globe, had not as yet destroyed all their Race? In effect, their life continutinued until the time of *Augustus*, according to the Testimony of *Plutarch*. It would even seem, that the prophetick and sacred Testament of our Primitive Patriarchs, designed to lead us by the Hand to that truth: For we read in it of the Revolt of Angels, before mention is made of Man. Is not that Sequel of time, which the Scripture observes, half of a Proof in a manner, that Angels inhabited the Earth before us? And that these proud Blades, who had lived in our World, whilst it was a Sun, disdaining, perhaps, since it was extinct, to abide any longer in it, and knowing that God had placed his Throne in the Sun, had the boldness to adventure to invade it? But God who resolved to punish their Audacity, banish'd them even from the Earth, and created Man less perfect, but by consequence less proud, to possess their vacant Habitations.

About the end of four Months Voyage, at least as near as one can reckon, when there is no Night to distinguish the Day; I came upon the Coast of one of those little Earths that wheel about the Sun, which the Mathematicians call Spots; where by reason that Clouds interposed, my Glasses now not uniting so much heat, and by consequence the Air not pushing my Shed with so much Force, what remained of the Wind could do no more, but bear up
my

my fall, and let me down upon the top of a very high Mountain, to which I gently descended.

I leave it to you to consider what Joy I felt, when I saw my Feet upon firm Ground, after I had so long acted the part of a Fowl. Words, indeed, are too weak to express the Extasie of Gladness I found my self in; when, at length, I perceived my Head Crowned with the Brightness of the Heavens. However, I was not so far transported yet with that Extasie, but that *I* thought of getting out of my Box, and of covering the Capital thereof with my Shirt before *I* left it; because I was apprehensive, that if the Air becoming Serene, the Sun should again kindle my Glasses, as it was likely enough, I might find my House no more.

By Gullies which seemed hollowed by the fall of Water, *I* descended into the Plain, where because of the thick Mud, that fatned the Earth, I had much ado to go: However, having advanced a little way, *I* arrived in a great Bottom, where *I* rencountred a little Man stark-naked, sitting and resting himself upon a Stone. *I* cannot call to mind whether *I* spoke to him first, or if it was he that put the Question to me: But it is as fresh in my Memory, as if *I* heard him still, that he discoursed to me three long Hours in a Language, which *I*

knew

knew very well *I* had never heard before, and which hath not the least resemblance with any of the Languages in this World; notwithstanding *I* comprehended it faster, and more intelligibly than my Mother Tongue. He told me, when *I* made enquiry about so wonderful a thing, that in Sciences there was a *true*, without which one was always far from the *easie*; that the more an Idiom was distant from this *truth*, the more it came short of the Conception, and was less easie to be understood. In the same manner, continued he, in Musick one never finds this *true*; but that the Soul immediately rises, and blindly aspires after it. We see it not, but we feel that Nature sees it; and without being able to conceive, in what manner we are swallowed up by it, it still ravishes us, tho we cannot observe where it is. *It's* the very same with Languages; he who hits upon that verity of Letters, Words, and Order in expressing himself, can never fall below his thought, he speaks always with congruity to his Conception; and it is because you are ignorant of this perfect *Idiom*, that you are at a stand, not knowing the Order, nor the Words, which might explain what you imagine. *I* told him, that the first Man of our World, had undoubtedly made use of that Language, because the several Names which he gave to several things, declared their Essence. He interrupted me, and went on. It is not absolutely necessary, for expressing all the mind conceives, but without it we cannot be understood of all. Seeing this Idiom is the Instinct or Voice of Nature; it ought to be intelligible to all that live under the Jurisdiction of Nature: And therefore if you understood it, you might Discourse and Communicate all your thoughts to Beasts, and the Beasts theirs to you; because it is the very Language of Nature, whereby she makes her self to be understood by all Living Creatures.

Be no more surprised, then, at the facility wherewith you understand the meaning of a Language, which never sounded before in your Ear. When I speak, your Soul finds in every Word of mine, that *Truth* which it gropes after; and though her Reason understand it not, yet she has Nature with her that cannot fail to understand it.

Ha! without doubt, cried *I*, it was by the means of that Emphatick Idiom, that our first Father heretofore conversed with Animals, and was by them understood; for seeing the Dominion over all the kinds of them, was given to him, they obeyed him, because he commanded in a Language that was known to them; and it is for

that

that Reason also, that (this Original Language being lost) they come not at present, when they are called, as heretofore they did, seeing now they do not understand us.

The little Man seemed as if he had no mind to answer me; but resuming his discourse, he was about to go on, if I had not once again interrupted him. I asked him, then, what World it was that we breathed in, if it was much inhabited; and what kind of Government they lived under? I am going, replyed he, to discover Secrets to you, which are not known in your Climate.

Consider well the Ground whereon we go; it is not long, since it was an indigested disorderly Mass, a Chaos of confused Matter, a black and glewy Filth, whereof the Sun had purged it self: Now after that by the force of the rays, which the Sun darted against it, he mingled, pressed, and compacted those numerous Clouds of Atomes: After, I say, that by a long and powerful Coction, he separated the more contrary, and reverted the more similary parts of this Bowl; the Mass pierced through with heat sweat so, that it made a Deluge, which covered it above Forty days; for so much Water required no less time, to fall down into the more declining and lower Regions of our Globe.

The

The Liquor of these Torrents being assembled, formed the Sea, which by its Salt makes it still apparent, that it must needs be a conflux of Sweat; all sweat being Salt. When the Waters were retired, a fat and fertile Mud remained upon the Earth: Now when the Sun shone out, there arose a kind of a Tumor or Wheal, which could not, because of the Cold, thrust out its bud: It therefore received another coction; and that coction still rectifying and perfecting it by a more exact mixture, it sent forth a Sprout; endowed then only with Vegetation, but capable of Sense: But because the Waters, which had so long stood upon the slime, had too much chilled it, the swelling broke not; so that the Sun recocted it once more; and after a third Digestion, that Matrix being so thoroughly heated, that the Cold brought forth a Man; who hath retained in the Liver, which is the seat of the vegetative Soul, and the place of the first Concoction, the power of Growing; in the Heart, which is the seat of Activity, and the place of second Concoction, the vital Power; and in the Brain, which is the seat of the Intellectual, and the place of the third Concoction, the power of Reasoning: Otherwise, why should we be longer in the Womb of our Mothers, than the rest of Animals, unless it be that our *Embryo* receives

ceives three diſtinct Concoctions, for forming the three diſtinct Faculties of our Soul; and the Beaſts only two, for forming their two Powers? I know that the Horſe is not compleated in the Belly of the Mare, before the tenth, twelfth, or fourteenth Month: But ſeeing he is of a Conſtitution, ſo contrary to that which makes us men, that he never has Life but in Months, which are obſerved to be fatal to ours, when we remain in the Womb beyond the natural Courſe; it is no wonder, that Nature needs another period of time for delivering a Mare, than that which brings a Woman to Bed. It is ſo; but in fine, ſome body may ſay, The Horſe remains longer than we in the Belly of his Mother; and by conſequence he receives there, either more perfect, or more numerous Coctions: I anſwer, that it follows not; for, not to rely upon the Obſervations, that ſo many Learned men have made upon the *Energy* of numbers, when they prove, That all Matter being in motion, ſome Beings are compleated in a certain Revolution of days, which are deſtroyed in another; nor yet to lay any great ſtreſs, upon the Arguments they deduce, from the Cauſe of all theſe motions, to prove that the number *Nine* is the moſt perfect; I ſhall content my ſelf with this anſwer, That the Bud of man being hotter, the Sun interferes

F and

and compleats more Organs in the space of nine Months, than he hath rough-hew'n in a Colt during a whole year. Now it is not to be doubted, but that a Horse is a great deal colder than a Man; seeing that Beast never dies, but of a Swelling of the Spleen, or other Diseases that proceed from Melancholy. Nevertheless, you'l tell me, there is no man in our World engendred of Mud, and produced in that manner: I believe it, your World at present is over-heated; for so soon as the Sun draws a sprout out of the Earth, finding none of that cold Humidity, or to say better, that certain Period of compleated Motion, which obliges it to several Coctions, it turns it presently into a Vegetable; or if it make two Coctions, seeing the second has not time enough to receive perfection in, it only engenders an Insect. And it is a Remark that I have made also, That the Ape, which, as we, carrys it's young almost nine Months, resembles us in so many Humors, that not a few Naturalists have ranked us in the same Species; and the reason is, that their Seed being of a temper, much like ours, hath during that time, had almost the leisure to perfect those three Digestions.

You'l undoubtedly ask me, of whom I have the Story, that now I have told you; you'l tell me, that I could not have had it from those that were not in being: It's true,

I

World of the Sun.

am the only person that hath hit upon it, nd by consequence I can give no Vouchers r it, because it's a thing that happened efore I was born; that's likewise true: But ke this along with you also, That in a Reion bordering upon the Sun, as ours does, e Souls full of Fire are more illuminated, ore subtile, and more penetrant, than ose of other Animals in remoter Spheres. ow seeing even in your World, there have en Prophets heretofore, whose minds, ightened by a vigorous Inspiration, have d Fore-knowledge of future things; it is t impossible, but that in this, which is far arer the Sun, & by consequence more luminus than yours, a strong *Genius* may have me smelling of what is past; that his acve Reason may move as well backwards, forwards, and that it may be able to atin to the Cause by the Effects, seeing it n reach the Effects by the Cause.

Thus he ended his Philosophical Discourse; but after a more particular Confence, that we had, about very deep Seets, which he revealed to me; part wheref I'll conceal, and of which the rest has caped me; he told me, That it was not as t three Weeks, since a clod of Earth, imegnated by the Sun, was brought to Bed him. Consider that Tumor attentively. hen he made me observe, I know not

F 2   what

what Swelling upon the Mud, not unlike to a Mole-Hill: That, says he, is an Apostume, or to say better, a Matrix, which for these Nine Months past, hath contained the Embryo of one of my Brothers. I wait here, on design to play the part of a Midwife to it.

He would have gone on, had he not perceived a Palpitation of the Earth, about that Swelling of Clay. That with the bigness of the Tumor, made him conclude, that the Earth was in Labour, and that that Shake, was already the effort of the Pangs of Travel. He thereupon immediately left me, that he might run to it; and for my part, I went to look for my Lodge.

I therefore clambered up again the Mountain, I had come down from, and was very weary before I got to the top of it. You may imagine what trouble I was in, when I did not find my House, where I had left it. I began to lament the loss of it, when I perceived it, skipping and vaulting at a great distance; I ran thither, as fast as my Legs could carry me, till I was out of Breath again; and really, it was an agreeable Diversion, to behold that new way of Coursing; for sometimes, when I had almost my Hand upon it, a new encrease of Heat, got into the Glass-Ball, which attracting the Air with more force, and that

Air

Air raising my Box up above my reach, made me leap after it, as a Cat does to a Hook, where it sees a Hare hanging. Had not my Shirt been upon the Capital, to resist the force of the Glasses, it would have made the Voyage by it self alone.

But to what purpose is it, to call to mind an Accident that I cannot think on, but with the same Grief, that I felt at that time! It's enough to know, that it bounded, rowed, and flew so long; and that I jumped, run, and skipped so much, that at length, I saw it fall down, at the Foot of a very high Hill. It had perhaps led me a Dance, a great deal farther, had not that lofty Swelling of the Earth, by its shade, that blacked the Sky, to a large distance upon the Plain, spread forth a Night of half a Leagues Circumference: For falling into that Darkness, the Glass thereof no sooner felt the Cool, but that no more Vacuity was engendred in it, no more Wind through the Hole, and by consequence, no more Impulsion to support it; so that it tumbled down, and would have broken into a Thousand shivers, had not by good luck, a Pond of Water, into which it fell, yielded under the weight. I drew it out of the Water, repaired what was bruised; and then having taken fast hold of it in my Arms, carried it to the top of a little Hill, hard by;

There I took my Shirt off on't; but I could not put it on, because my Glasses beginning to work their effect, I perceived my Castle already wagging to be gone. I had no more time left, but suddenly to get in, where I shut my self up as at first.

The Sphere of our World, appeared to me as a Planet, much about the bigness of the Moon, as she appears to us: Nay, it grew less and less, still proportionably as I ascended, insomuch that it became a Star, then a Spark, and then nothing at all; for that luminous Point grew so sharp, that it might equal that, which terminates the last Ray of my sight; that at length it was swallowed up in the common colour of the Heavens. Some perhaps may wonder, that during so long a Voyage, I was not overtaken with Sleep; but seeing Sleep is only produced by the gentle Exhalation of the Victuals, which evaporate from the Stomach to the Brain, or by a Necessity that Nature finds of locking up our Soul, that during Repose, she may recover as many Spirits, as have been spent in Exercise: I had no mind to Sleep, since I did not Eat, and that the Sun supplied me, with more radical Heat, than I consumed. In the mean while, my Elevation continued, and proportionably as it brought me nearer to that enflamed World, I felt a certain Joy diffuse

it

it self through my Blood, which rectified it, and advanced to my very Soul. Ever now and then I looked upwards, that I might admire the Vivacity of the Lustre, that rayed into my little Chrystal-House; and it is fresh in my Memory still, that at the time I put my Eye to the Orifice of the Vessel, behold, with a sudden start, I felt something heavy, that fled from all the parts of my Body. A whirling Cloud of very thick, and, in a manner, palpable Smoak, choaked my Glass with Darkness; and when I stood up to contemplate that Obscurity, which blinded me, I saw no more, neither Vessel, nor Burning-Glasses, Glass-Window, nor Covering to my Shed: I looked down, then, with Design to see what made my Master-piece thus fall to ruine; but in place of it, and of the four Sides and Floor, I found nothing but the Heavens round about me. I was frightned beyond measure, when I found, as if the Air had been petrified, I know not what invisible Obstacle, which forced back my Arms, when I thought to stretch them forth. It came then into my Imagination, that mounting so high, I had without doubt got into the Firmament, which some Philosophers and Astronomers have maintained to be solid. I began to fear, I should remain studded in there; but the Horrour that the

odd-

oddness of that Accident cast me into, was exceedingly encreased by those that followed: For my sight, that rambled here and there, falling by chance upon my Breast, instead of stopping at the Surface of my Body, went quite through; then a moment after, I thought I saw behind me, and almost without any interval, as if my Body had been no more but an Organ of sight: I perceived that my Flesh, being cleansed from its Opacity, transferr'd Objects to my Eyes, and my Eyes Objects through it. At length, having above a Thousand times knockt against the Roof, Floor, and Walls of my Chair-volant, without being able to see it; I found, that through a secret Necessity of the Light in its source, my Shed and I were both become transparent. Not but that I ought to have perceiv'd it, though *Diaphanous*, seeing we very well perceive Glass, Chrystal, and Diamonds that are so; but I fancy that the Sun, in a Region so near himself, purges Bodies more perfectly from their Opacity, by ranking more straightly the imperceptible Pores of the Matter, than in our World; where his Force, worn out almost by so long a passage, is hardly able to transfuse his Lustre into precious Stones: However, by reason of the internal Smoothness of their Surfaces, he makes them reflect through their Glasses, as through little Eyes,

Eyes, either the Green of Emeralds, the Scarlet of Rubies, or the Violet of Amethyſts, according as the different Pores of the Stone, either more ſtraight or more winding, extinguiſh or rekindle that weak Light, by a great many Reflexions. One difficulty may puzzle the Reader, to wit, How I could ſee my ſelf, and not my Houſe, ſeeing I was become Diaphanous, as well as it? I anſwer, That without doubt the Sun acts otherwiſe upon animate, than upon inanimate Bodies; becauſe no part neither of my Fleſh, Bones nor Entrails, tho' tranſparent, had loſt its Natural Colour; on the contrary, my Lungs under an incarnate Red, ſtill retained their delicate Softneſs: My Heart ſtill Vermilion, gently moved with its *Syſtole*, and *Diaſtole*: My Liver ſeemed to burn in a Purple Red, and boyling the Air which I breathed, continued the Circulation of the Blood: In ſhort, I ſaw, touched, and felt my ſelf to be the ſame, and nevertheleſs I was not ſo.

Whilſt I was conſidering this Metamorphoſis, my Voyage grew ſhorter and ſhorter ſtill, but with a great deal of ſlowneſs then, by reaſon of the Serenity of the *Æther*, which was rarified proportionably, as I drew nearer the ſource of Day; for ſeeing the matter in that Region is very thin, becauſe of the great Vacuity it is full of, and that by conſequence

that

that Matter is very lazy, by reason of the Vacuity that has no *Action*, that Air passing through the hole of my Box, could not produce but a little Wind, hardly able to support it.

I never reflected upon the Malicious Capriciousness of Fortune, which always so obstinatly opposed the Success of my undertaking, but I wonder my brains did not turn. But listen to a Miracle, which future Ages will hardly be inclined to believe.

Being shut up in a Box as clear as day, that I had just lost sight of, and my flight flagging so, that I had much ado not to fall; in a word, being in a condition, that all that's contained in the great Fabrick of the World, was unable to assist me; I found my self reduced to the Period of extream Misfortune: Nevertheless, as when we are expiring, we find an internal Impulse in us, to embrace those who have given us a Being; I lifted up my Eyes to the Sun, our common Father. That ardour of Will, not only supported my Body, but also raised it up towards the thing, which it aspired to embrace. My Body pushed forwards my Box, and in that manner I continued my Voyage. So soon as I perceived this, more intensly than ever, I plyed all the faculties of my Soul, to raise my Imagination towards that which attracted me; but my head being loaded with my Shed, against the upper-part whereof, the Efforts

of

of my Will pusht it, whether I would or not, that did so incommode me, that at length so much weight, forced me to grope for the place of its invisible Door. By good fortune I found it, and having opened it, threw my self out: But that natural Apprehension of falling, which all Animals have, when they find nothing to support them, made me briskly stretch forth my Arm, that I might take hold of somewhat: I had no other Guide but Nature, which stands not upon Reasoning; and therefore Fortune, her Enemy, maliciously forced my hand upon the Capital of Chrystal. Alas! what Thunderclap to my Ears, was the sound of the Icosaedron, which to my hearing broke in pieces! Such a Disorder, Misfortune, and Fright are beyond all expressions! The Glasses attracted no more Air, for no more Vacuity was made; the Air became no more wind, by its hastening to fill it, and the wind ceased to carry my Box on high: In short, immediately after that wrack, I saw it long a falling through those vast plains of the World. It recontracted in the same Region, the dark Opacity which it had exhaled; in respect that the energetick Vertue of the Light, ceasing in that place, it greedily vnited again to the obscure Thickness, which was in a manner essential to it; in the same manner as Spirits, long after their Separation on, have been seen to come in search of their Bodies, and that they might rejoyn them, to wander for the space of an hundred years about their Graves. I fancy it lost in this manner its Transparency, for I have seen it since in *Poland*, in the same condition it was in, when first I entered it. Now I have been informed, that it fell under the Equinoctial Line, in the Kingdom of *Borneo*; that a *Portuguess* Merchant bought it of the Islander that found it, and that from hand to hand, it fell into the possession of that *Polish* Engineer, who makes use of it at present to fly with.

Thus, then, hanging in the open space of the Heavens, and terrified already by the Death, which my fall threatned me with, I turned, as I told you, my sad eyes towards the Sun: My Sight carried my thought thither, and my Looks being fixed upon his Globe, marked out a way which my Will followed, to transport my Body to the same place.

That vigorous Launching out of my Soul, will not be incomprehensible to any, that will but consider, the simple Efforts of our Will; as, for Instance, It is very well known, that when I have a mind to leap, my Will being excited by my Fancy, raises the whole Microcosm, and endeavours to transport it to the mark, that it

pro-

proposed to it self. If it attain not always to it, it's because that the Universal Principles of Nature prevail over the Particular; and that the Power of Willing being particular to sensitive Beings; and that of falling to the Center, proper to all things, mater in general; my Leap is forced to cease, so soon as the Mass of my Body, having overcome the Insolence of the Will, that surprized it, draws near the Point to which it tends.

I shall wave what else happened to me, during the rest of my Voyage, lest I should be as long in relating, as I was in making it: I'll only tell you in general, that at the end of 22. Months, I at length happily arrived at the great plains of Day.

That Land looks like flakes of burning Snow, so luminous it is; nevertheless it is a thing pretty incredible, that I could never comprehend, after that my Box fell, whether I mounted up, or descended towards the Sun. I only remember, when I arrived there, that I walkt lightly; I toucht only the Ground in a point, and I often rowled like a Bowl, finding it alike commodious, to walk either upon my Head or Feet. Though sometimes my Feet were towards Heaven, and my Shoulders towards the Earth; yet in that posture, I found my self as naturally situated, as if my Feet had

been

been towards the Earth, and my Shoulders towards Heaven. Upon what part soever of my Body I placed my self, whether upon my Belly, or Back, on Elbow or Ear, I found my self standing. By that I knew, that the Sun is a World, which hath no Center; and that as I was far enough from the active Sphere of ours, and of all the others which I met; so by consequence, it was impossible, that I should still be ponderous, seeing Weight is nothing else, but an Attraction of the Center, within the Sphere of its Activity.

The Respect, wherewith I printed my steps upon that Luminous Plain, suspended for some time, the eager Desire I had to advance on my Journey. I was all ashamed to walk upon the Day: Nay, and my astonished Body, desiring to relie upon mine Eyes; and that transparent Ground, which they pierced, not being able to support them; my Instinct in spight of me, now become Master of my thought, hurried it into the Abyss of a bottomless Light. My Reason nevertheless, by degrees undeceived my Instinct; I walked confidently, and without trembling upon the Plain, and directed my steps so boldly, that if Men could have perceived me from their World, they would have taken me for some Power marching upon the Clouds. Having, as I think,

tra-

travelled about fifteen days time, I came into a Country of the Sun, not so resplendent as those I had left. I found my self transported with Joy, and I imagined, that undoubtedly that Joy proceeded from a secret Sympathy, which my Being still retained with its Opacity. Nevertheless, the knowledge I had of this, made me not desist from my Enterprise; for then I was like to those sleepy Old men, who tho' they know that sleep is hurtful to them, and that they have ordered their Servants, to snatch them out of it's Clutches, are nevertheless very angry when they awaken them. In like manner, tho' my Body growing obscure, as I reached the darker Provinces, recontracted the weaknesses, that that infirmity of matter brings along with it; yet I was pleased therewith: Growing weary, sleep invaded me; for that flattering Heaviness, wherewith the approaches of sleep charms us, distilled so much pleasure into my Veins, that my Senses overcome thereby, forced my Spirit to congratulate the Tyrant, who chained its Servants; for Sleep, that Antient Tyrant of one half of our days, who because of his Old age, being unable to support the Light, or to look on it without fainting, was forced to forsake me, upon my entry into the shining Countries of the Sun, was come to expect me on the Confines of the dusky Region I speak of; where having caught

caught me, he made me Prisoner, shut up my Eyes, his declared enemies, under the dark vault of my Eye-lids; and for fear that my other Senses betraying him, as they had betrayed me, might disturb him in the peaceable possession of his Conquest, he tied them fast to their several Beds. The meaning of all this is in two words, that I lay down upon the Sand, and fell asleep. It was a plain Country, and so open that as far as my sight could reach, I did not discover so much as one Bush; and nevertheless when I awoke, I found my self under a Tree, in respect of which the tallest Cedars would but appear as Grass. The Trunk of it was of Massive Gold, it's Branches of Silver, and it's Leaves of Emeralds, which upon the resplendent Verdure of their precious Surface, represented, as in a Looking-Glass, the Images of the Fruit that hang about them. But judge ye whether the Fruit owed any thing to the Leaves; the enflamed Scarlet of a large Carbuncle, composed one half of every one of them; and the other was in suspence, whether it held it's matter of a Chrysolite, or of a piece of gilt Amber; the blown Blossoms were large Roses of Diamonds, and oriental Pearls the Buds.

A Nightingale, whose smooth Plume rendered it exceeding lovely, perching on the highest sprig, seemed sollicitous, with it's

Me-

Melody, to force the Eyes to confess to the Ears, that it well deserved the Throne whereon it sate.

I stood a long while amazed at the sight of that rich Spectacle, and I could not have my full of beholding it: But whilst my thoughts were wholly taken up in contemplating, amongst the other Fruit, an exceeding lovely Pomegranate, whose Flesh was a swarm of large Rubies in clusters, I saw that little Crown that stands it instead of a Head move, which lengthened it self, as much as was needful, to form a Neck; next I saw somewhat white bubble and boil over it, which by means of Condensation, Concretion advancing and repelling the matter in certain places, appeared at length to be the face of a little bulk of Flesh. That little bulk shaped it self into a round Figure towards the girdle, that's to say, that in the lower part of it, it still retained the shape of an Apple. Nevertheless it extended it self by degrees, and the tail of it being converted into two Legs, each Leg divided it self into five Toes. So soon as the Pomegranate was humanized, it broke off from its stalk, and with a gentle Toss fell just at my Feet. I confess, really, when I saw marching stately before me that rational Apple, that little Butt-end of a Dwarf, no bigger than my Thumb, and yet so powerful as to create himself; I was seized

with

with Veneration. Human Animal (said he to me, in that Original Tongue I told you of before,) after I had long considered thee, from the top of the Branch where I hung; I thought I read in thy countenance, that thou wast no Native of this World, and that's the reason why I am come down to be informed of the truth. When I had satisfied his Curiosity, as to all the questions he put to me——But pray, said I to him, tell me who you are; for what I have now seen, is so strangely surprising, that I despair of ever knowing the Cause of it, unless you discover it to me. How! a great Tree all of pure Gold, the Leaves whereof are Emeralds, the Flowers Diamonds, the Buds Pearls; and besides all that, Fruit that make themselves men in the twinkling of an Eye? For my part, I confess it passes my Capacity, to comprehend such a Miracle. Having uttered this Exclamation, whilst I expected his answer: You will not take it amiss, said he to me, if being King of all the people that make up this Tree, I call them to follow me. When he had so said, I observed that he recoiled within himself: I cannot tell, whether by bending the internal springs of his Will, he excited without him some Motion, that produced what now you shall hear: But so it is, That immediately after, all the Leaves and Branches, in short the whole Tree, broke

to

to pieces, and became little Men, that saw, felt, and walked, who as if they intended to celebrate their Birth-day, at the very instant of their production, fell a dancing about me: Of all, I saw none but the Nightingale retained it's former shape, and was not at all Metamorphosed; it came and perched upon the Shoulder of our little Monarch, where it Sung so melancholick and amorous an Air, that the whole Assembly, and the Prince himself, mollified by the sweet Languishings of its dying Voice, could not forbear to shed some Tears. A Curiosity to learn whence that Bird came, seized me at that time, with such an extraordinary Itch of Tongue, that I could not for my heart restrain it. Sir, said I, addressing my self to the King, did I not fear to trouble your Majesty, I would ask you the question, why amongst so many Transformations, the Nightingale alone hath retained its kind? That little Prince listened to me with so much Civility, that it shew'd he had a great deal of good Nature; and knowing my Curiosity: The Nightingale, replied he, hath not changed its Form as we have done, because it could not; it's a real Bird, and nothing else than what it appears to be to you. But let's go towards the obscurer Regions, and by the way I'll tell you who I am, and give you the Story of the Nightingale. Hardly had I intimated

ted to him, the Satisfaction that I receiv'd from his offer, when he skipped nimbly up upon one of my Shoulders. He stood upon his little Tip-toes, that he might reach my ear with his Mouth ; and sometimes swinging, sometimes pestered in my Hair: In troth, said he, thou must e'en excuse one that's already out of breath ; seeing in a narrow Body my *Lungs* are contracted, and by consequence my Voice so small, that I am forced to take a great deal of pains, to make my self be heard : The Nightingale would do best, to tell it's own History it self ; let it sing, then, if it think fit, we shall have the Pleasure, at least to hear it's Story in Musick. I replied, that I was not as yet, sufficiently acquainted with the Language of the Bird ; that a certain Philosopher, indeed, whom I met with, as I was mounting up to the Sun, had given me some general Principles, for understanding the Language of Brutes ; but that they were not sufficient, for the understanding of all words in general, nor for being affected with all the Delicacies that occur in an adventure, such as that must needs be. Well then, said he, since thou'lt have it so, thine Ears shall not only be deprived of the sweet Songs of the Nightingale ; but in a manner of all its adventure also, of which I can tell thee no more, than what is come to my knowledge : However be content with that Pattern ; besides tho'

I

I knew it fully, yet the shortness of our journey into it's Country, whither I am going to carry it back again, would not suffer me to enlarge in my Relation. Having said so, he jumpt down from my Shoulder upon the ground; then he gave the hand to all his little People, and fell a dancing with them in a kind of Figure, that I cannot represent, the like having never seen. But hearken, O ye people of the Earth, to what I don't at all oblige you to believe; seeing in that World, where your Miracles are but natural Effects, this hath past for a real Miracle. So soon as these little men were fallen a dancing, I thought I felt their Agitation in my self, and my Agitation in them. I could not behold that Dance, but that I was sensibly drawn out of my place, as by a *Vortex* that moved by it's own whirling, and by the particular Agitation of every one of them, all the parts of my Body were put into Agitation; so that I felt the same Gayness flush in my countenance, which a like Motion had spread upon theirs. As the Dance closed more and more, the Dancers shuffled together, by a prompter and more imperceptible way of skipping. The design of the Ball seemed to be, to represent a huge Giant; for by approaching to one another, and redoubling the swiftness of their Motions, they mingled so close together, that I distinguished no more, but a great transparent *Colossus*; however my eyes saw them enter one into another. About that time, it began to be out of my power, any more to discern the Diversity of their several Motions, by reason of their extream Volubility; and also because that Volubility, contracting more and more, according as it approached to the Center, each vortical Trip at length, took up so little a space, that it escaped my eye-sight. Nevertheless I believe, that the parts still approached one another; for that humane Mass, heretofore so hugely great, was by degrees reduced into the shape of a young Man, of a middle stature, whose Members were proportioned with so exact a Symetry, that the highest Idea of perfection could never reach it: He was beautiful, beyond all that the fancy of Painters could ever imagin; but that which seemed to me very strange was, that all the parts which compleated that perfect Microcosm, were linked together in the twinkling of an eye; the nimbler sort of our little Dancers cut a Capriol, to the height and natural posture of forming a Head. The hotter, but less agile, formed the Heart; and those that were much heavier, furnished only the Bones, Flesh, and Plumpness.

When that lovely big young Man, was compleatly finished, though the speedy Com-

Compofition of him, had hardly allowed me time to obferve any interval in the Progreffion; I faw the King of all thefe People, of whom he was an Abridgement, enter in at the Mouth: Nay, it feemed to me, that he was drawn into that Body, by the Refpiration of the Body it felf. This Pile of little Men, had not as yet given any fign of Life; but fo foon as it had fucked in its little King, it perceived it felf then, to be no more but one. For fome time he ftood confidering of me, and feeming by his looks, now acquainted with me; he drew near, careffed me, and giving me his Hand: Now it is, faid he, that without wronging the delicatenefs of my Lungs, I can entertain thee with the things, that thou paffionately defiredft to know: But firft of all, it's but reafonable, that I fhould difcover to thee, the hidden Secrets of our Original. Know then, that we are native Animals of the brighter Regions of the Sun; the moft ordinary, as well as the moft ufeful of our Employments, is to Travel over the vaft Countries of this great World. We curioufly obferve the Manners of People, the Genius of Climates, and the Nature of all things, that deferve our Confideration; by means whereof, we form to our felves, a certain Science of every thing that hath a Being. Now thou muft know, that my Vafsals travelled under my Conduct; and to the end, we might have time to observe things more curiously, we retained not that particular Conformation of our Bodies, which cannot fall under thy Senses, and whose Subtilty would have made us make too much haste; but we converted our selves into Birds; all my Subjects, by my Orders, became Eagles; and as for my self, least they might grow weary, I Metamorphosed my self into a Nightingale, to sweeten their Labour, by the Charms of Musick: Without flying, I followed the rapid flight of my People; for I perched upon the Head of one of my Vassals, and we were still pursuing our Journey; when a Nightingale, an Inhabitant of a Province of the obscure Country, that at that time we Travelled through, astonished to see me in the Power of an Eagle (for it could take us for none other, but what it saw us to be) fell a bewailing my Misfortune: I ordered my People to halt, and we lighted on the top of some Trees, where that Charitable Bird lamented. I took so much Pleasure in the sweetness of her doleful Airs, that to the end I might enjoy them the longer, and with more convenience, I resolved not to undeceive her. I devised upon the spot a Story, wherein I told her the imaginary Misfortunes, that had made me fall into the power of that Eagle: I pieced it out with

so

so surprising Adventures, which so dexterously raised the Passions, and the Air was so well adapted to the Letter, that the Nightingale, was quite beside her self. We reciprocally warbled to one another, the History of our mutual Love in Musick. I chanted in my Airs, that not only I was comforted, but even rejoyced in my disaster, seeing it had procured me the Glory, to be lamented in such pretty Songs; and that little disconsolate Bird answered me in hers, that she would joyfully accept all the esteem, I put upon her; if she thought, that that could make her deserve, the Honour of dying in my place; but that Fortune, not having reserved so much Glory, for such a Wretch as she, she only accepted of that esteem, as much as might hinder me, from being ashamed of my Friendship. I made answer again, on my part, with all the transports, tenderness, and softness of so touching a Passion, that I perceiv'd her, three or four times, ready to die of Love, upon the Branch. The Truth is, I mingled so much Art, with the Sweetness of my Voice, and surprized her Ear with such quaint Touches, and by ways so unusual to those of her Kind, that I raised in her pretty Soul, what Passion soever I pleased. In this Exercise, we spent four and twenty Hours; and I believe, we had never given over making of Love, had not our Throats denied us any more Voice. That was the only Obstacle, that hindred us from proceeding: For perceiving, that the Pains I took, began to tear my Throat, and that I could hold out no longer, without falling into a Swoon, I made her a sign to draw near to me. The danger she thought me to be in, amidst so many Eagles, perswaded her, that I called her to my aid: She came flie-

ing

ing immediately, to my Assistance, and resolving to give me a Glorious Instance, that she could for a Friend brave Death even upon his Throne, she boldly lighted upon the great crooked Beak of the Eagle, where I was perched. Really, so strong a Courage, in so weak a Creature, affected me, with some Veneration; for, grant I had implored her aid, as she fancied, and that it be a Law, amongst Animals of the same kind, to assist the unfortunate; yet, the Instinct of her timorous Nature, ought to have made her waver; and nevertheless, she boggled not in the least: On the contrary, she made so much haste, that I cannot tell which flew first, the Signal, or the Nightingale. Proud, to see under her Feet, the Head of her Tyrant, and ravished, to think that for my sake, she was to be Sacrificed, almost under my Wings, and that some happy drops of her Blood, might perhaps Jert upon my Feathers, she gently turned her Eyes to me, and having bid me *adieu*, as it were, by a Glance, which seemed to ask me leave to die; she struck so briskly her little Beak, into the Eyes of the Eagle, that they seemed to me to be out, before the peck was given. When my Bird perceived it self to be blind, it formed to it self another fight of new. I gently rebuked the Nightingale, for her too rash Action; and thinking it would be dangerous to conceal our real Being, any longer from her, I told her, who we were; but the poor little thing, prepossessed with an Opinion, that these Barbarians, whose Prisoner I was, forced me to devise that Tale, gave no credit to all that I could say to her. When I found that all the Reasons, whereby I thought to convince her, proved ineffectual

fectual, I gave private Orders, to ten or twelve thousand of my Subjects, and immediately, the Nightingale, perceived under her Feet, a River, running under a Boat, and the Boat floating upon it; the Boat was no bigger, than was sufficient to hold me, and another of my Size. At the first Signal given, my Eagles flew away, and I threw my self into the Skiff; from whence, I called to the Nightingale, that if she could not as yet, resolve to leave me so soon, she should embark with me. So soon as she was come in, I commanded the River to take its course, towards the Region, whither my People flew; but the fluidity of the Water, being inferiour to that of the Air, and by consequence, the Rapidity of their flight greater than that of ours, we were left a little behind. During the whole Voyage, I made it my Business, to undeceive my little Passenger; I told her, that she ought not to expect any fruit of her Passion, since we were not of the same Kind, that she might very well have perceived that, when the Eagle, whose Eyes she had struck out, framed to it self new ones, in her presence, and when at my command, twelve thousand of my Subjects, had Metamorphosed themselves, into that River and Boat, which carried us. My Remonstrances had not the least success: She made me answer, that as for the Eagle, who, I would have it believed, had formed to it self Eyes, it had no need of them; because he had not struck her Beak right into the Ball of its Eye; and as to the River and Boat, which I said to have been begot only of a Metamorphosis of my People, they were in the Wood from the Creation of the World, though they had not

been

been minded. Perceiving her so Ingenious, in deceiving her self, I agreed with her, that my Vassals, and I, should Metamorphose our selves to her view, into what she pleased; provided, that after that, she would return to her own Country. Sometime, she desired it should be into a Tree; sometime, she wished it might be into a Flower; sometime into Fruit, sometime into Metal, and sometime into Stone. In fine, that I might at once satisfie all her Desires, when we arrived at my Court, where I ordered her to expect me, we Metamorphosed our selves, to the Eyes of the Nightingale, into that precious Tree, thou foundest upon the Road, of which we have just now abandoned the form. Now after all, that I see that little Bird, resolved to return into her own Country, my Subjects, and I, are about to resume our Figure, and the right way of our Journey: But it is but reasonable, that I should first discover to thee, that we are Natives, and *Aborigenes* of the Sun, in the bright part thereof; for there is a very remarkable Difference, betwixt the People, which the Luminous Region produces, and the People of the obscure Country. We are they, whom in the World of the Earth, ye call Spirits, and your presumptuous stupidity, hath given us that Name; because, imagining no Animal, more perfect than Man, and perceiving that some Creatures, perform things above Humane Power, you have taken these Animals for Spirits. You are mistaken though, we are Animals as well as you: For although when we please, we give to our Matter, as you have just now seen, the essential Figure, and Form of the things, into which we have a mind to transform our selves,

that

that does not infer that we are Spirits. But liften, and I'll difcover to thee, how all thefe Transformations, which feem to thee to be fo many Miracles, are no more but pure natural Effects. Thou muft know, that being born Inhabitants of the bright part of this great World, where it is the Principle of Matter to be in Action; we ought to have the Imagination, far more active, than thofe of the obfcure Regions, and the Subftance of Body, far more fubtil alfo. Now this being fuppofed, it muft needs be, that our Imagination meeting with no Obftacle, in the matter that compofes us, it difpofes the fame as it pleafes, and becoming Miftrefs of all our Mafs, makes it by moving all its Particles, to pafs into the order, neceffary for conftituting that great thing, which it had formed in little. So that every one of us, having imagined the place, and part of that precious Tree, into which we had a mind to be changed; and having by that effort of Imagination, excited our matter to the Motions, neceffary for producing them, we transformed our felves into the fame. Thus my Eagle, having his Eyes peckt out, had no more to do, for reftoring of them again, but to imagine himfelf a fharp-fighted Eagle; for all our Transformations, are performed by Motion; and therefore it is that, when from Leaves, Flowers, and Fruit that we were, we became tranfmuted into Men, thou faweft us dance ftill fometime after, becaufe we were not as yet recovered from the agitation, which we ought to give to our matter, for making of us Men: After the manner of Bells; which though they be ftopt, yet Chime for fome time after, and dully retain the fame

found

sound, which was caused by the striking of the Clapper; and therefore thou sawest us dance, before we made that great Man, because for production of him, it behoved us to take all the general, and particular Motions, that were necessary to constitute him; to the end that that agitation by degrees compressing, and compacting our several Bodies by it's motion, might create in every part the specifick Motion, which it ought to have. Ye men of the other world, cannot do the same things, by reason of the Heaviness of your Mass, and the Coldness of your Imagination.

He continued his Probation, and illustrated it by so familiar and palpable Instances, that at length I was undeceiv'd of a great many ill prov'd opinions, wherewith our obstinate Doctors possess the minds of the weak. At that time I began to conceive, that in reality, the imagination of these Solar people, which by reason of the Climate ought to be hotter, their Bodies for the same reason lighter, and their persons more active (there being in that World, as there is in ours, no Activity of the Center, which may divert the matter from the motion, which that Imagination stamps upon it:) I conceived, I say, that that Imagination might produce without a Miracle, all the Miracles which it had performed. A thousand Examples of almost the like effects, affirmed by people of our Globe, fully perswaded me of this. *Cippus* K. of *Italy*, who having been present at a fight of Bulls, and his imagination all the night after, running upon Horns, found his forehead horned next morning. *Gallus Vitius*, who bent his mind, and so vigorously excited it, to conceive the Nature of Folly; that having by an effort of Imagination, given to his matter the same motions, that that matter ought to have for constituting Folly, became a Fool. King *Codrus* being Ptisical, and fixing his eyes and thoughts upon the fresh looks of a young Countenance, and that florid Chearfulness which darted upon him, from the youthfulness of the Boy, exciting in his Body the motion, whereby he fancied the healthfulness of a young man, recovered his Health. In short, many Women have made their Children, already formed in the Womb, Monsters; because their imagination, that was not strong enough, to give to themselves the Figure of the Monsters, which they conceived, had force sufficient to muster the matter of the *Fætus*, that was much hotter, and more moveable than their own, into an order propper for the production of these Monsters. Nay I was perswaded, that when that famous Hypochondriacal man of Antiquity, imagined himself to be a Pitcher, if his two compact and heavy matter could have followed the emotion of his Fancy, it would have formed of his entire Body a perfect Pitcher; and he would have appeared to all men to be a real Pitcher, so as he appeared to be to himself alone. A great many other Instances, wherewith I satisfied my self, so throughly convinced me, that I made no doubt any more, of the wonders which the Man-spirit told me. He asked me, if I desired any thing else of him; and I thanked him with all my Heart: After that, he had still the goodness to advise me, that since I was an Inhabitant of the Earth, I should follow the Nightingale into the obscure Regions of the Sun; because they were more sutable to the pleasures which Human Nature covets. No sooner had he concludcluded this discourse, but that opening his Mouth very wide, I saw the King of these little Animals come out of his Throat, in shape of a Nightingale. The great man instantly fell down, and at the same time, all his Members by morcels flew away under the form of Eagles. That Nightingale self-Creator, perched upon the fairest of them, from whence he warbled out a most excellent Air, whereby, I fancy, he gave me the Farewel. The real Nightingale took flight also, but not the way as they did, nor did she soar so high; so that I did not lose sight of her. We travelled much about the same rate; for seeing I was indifferent what Country I went to first, I was very willing to accompany her; besides that the obscure regions of the Birds, being more conform to my Constitution, I hoped also to meet with Adventures there, more agreeable to my humour. In that Expectation I travelled for the space of three weeks at least, with all imaginable content, had I had nothing but my ears to satisfie; for the Nightingale let me not be without Musick; when she was weary, she came and rested upon my Shoulder; and when I stopt, she staid for me. At length I arrived in a Country, of the Kingdom of that little Quirister; who then cared no more for my Company, so that I lost sight of her. I sought her, and called to her; but at length growing weary of running up and down in vain after her, I resolved to take some rest: For that end I laid my self along upon a soft Carpet of Grass, spread at the Root of a lofty Rock, that was covered with many trees, whose blith and fresh Verdure was a perfect Emblem of Youth: But whilst softened by the Charms of the place:

The

The History of BIRDS.

I Began to fall asleep in the Shade, I perceived in the Air a strange Bird, that hovered over my Head; it supported it self by so slight and imperceptible a motion, that I was many times in doubt, whether it might not be also a little Universe, balanced by its own Center. However by little and little it descended, and at length came so near, that it filled my Eyes with a delightful Prospect. The Tail of it seemed to be green, its Breast Azure-enamel'd, its Wings Incarnate, and its Head Purple, which tossed a glittering Crown of Gold, the Rayes whereof sparkled from its Eyes.

It kept a long time upon the Wing, and I was so attentive to observe what became on't, that my Soul being contracted, and in a manner wrapt up in the sole action of Seeing, it hardly reached my Ear, to let me hear that the Bird spoke as it sung.

However, being by little and little unbent from my Extasie, I distinctly remarked the Syllables, Words, and Discourse which it uttered.

To the best of my Memory, then it spun out its Song into these terms,

You are a Stranger, whistled the Bird, and has had your birth in a World, of which originally I am. Now that secret propensity to mutual Love, that those of the same Country have one for another, is the instinct, which Inclines me to inform you of my Life.

I perceive your mind puzled to conceive, how it is possible that I should express my self to you, in a continued Discourse, seeing though Birds imitate your words, they understand not the meaning of them; but, on the other hand, when you counterfeit the Barking of a *Dog*, or the Note of a *Nightingale*, you understand as little what the *Dog* or *Nightingale* means thereby. Conclude then from thence, That neither Birds nor Men are therefore the less rational.

Nevertheless, in the same manner as amongst you, there have been some so Ingenious, as to have understood and spoken our Language, as *Apollonius, Thianeus, Aneximander, Æsop*, and many others, whose names I conceal, because they never came to our knowledge; so there are those amongst us, who understand and speak yours. Some indeed, know only the Language of one Nation: But just so

as

as there are some Birds that say nothing at all, some that chirp, and others that speak; there are also some more perfect, who can make use of all Idioms; as for my self, I have the honour to be one of that small number.

After all, you must know, that in what world soever it be, nature hath stamped on Birds a secret desire of flying up hither; and perhaps that Emotion of our Will is, that which hath made our Wings to grow; as Women with Child imprint upon their Children, the figure of the things they have longed for; or rather as those who passionately would learn to Swim, have been seen in their sleep to jump into Rivers, and with more skill than an experienced Swimmer pass those dangers, that if they had been awake, they durst not have so much as beheld; or as the Son of that same King *Cræsus*, whom a vehement desire of speaking, to save his Father's Life, taught a Language all of a sudden; or in short, as that Ancient who being pursued by his Enemy, and surprised without Arms, had Bulls Horns grow out on his Fore-head, through the desire that a Rage, not unlike to that of that Beast, inspired into him.

When Birds then arrive in the Sun, they go and associate themselves to the Republick

lick of their kind. I well perceive, you are big with expectation to learn what I am, it is I who amongst you am called a *Phenix*, in every world there is but one at a time, which lives there for the space of an hundred Years; for at the end of an Age, when upon some Mountain of *Arabia*, it hath laid a great Egg amidst the Coals of its Funeral Pile, which it hath made of the Branches of Aloes, Cinnamon and Frankincense, it takes its flight, and diverts its course towards the Sun, as the Country to which its heart hath long aspired. It hath indeed made many Attempts before, for accomplishing that Voyage; but the weight of its Egg, which hath so thick a shell, that it requires an Age to be hatched in, still retarded the Enterprise.

I am sensible, that you can hardly comprehend that miraculous Production; and therefore I'll explain it to you. The *Phenix* is an Hermaprodite, but amongst Hermaphrodites, it is likewise another *Phenix* altogether extraordinary: For––––––––

It continued half a quarter of an hour without speaking, and then added: I perceive you suspect what I have told you to be false, but if what I say be not true, the first time I come into your Globe, may an Eagle devour me.

It

It remained a little while longer hovering upon Wing, and then flew away.

The Admiration that its relation put me into, gave me the Curiosity to follow it; and because it cut the Welkin with no swift flight, I easily kept within sight of it.

At the end of Fifty Leagues, or thereabouts, I found my self in a Country so full of Birds, that their number equalled almost the number of the leaves that covered them. The thing that surprised me most was, that in stead of growing wilder upon my coming, they flew about me, one chirped into my Ears, another fetched a compass about my head: In short, when their Gambols had for a long time taken up my attention, all of a sudden I felt my Arms loaded with above a Million of all sorts and kinds, which pressed them down so heavily, that I could not move them.

They held me in this posture, until I saw Four great Eagles come, of which Two having with their Talons taken fast hold on me by the Legs, and the other Two by the Arms, they carried me up aloft in the Air.

Amongst the Wood I spied a Magpy, that made a great bustle flying up and down; and I heard her call to me, that I should

should not make resistance, because her Companions were already consulting to put out my Eyes. This admonition put a stop to all the Strugling that I could have made; so that these Eagles carried me above a Thousand Leagues from thence into a great Wood, which was (as the Magpy told me) the City where their King held his Residence.

The first thing they did, was to throw me into Prison, in the hollow Trunk of a large Oak; and a great many of the strongest perched upon the Branches, where they performed the duty of a Company of Souldiers standing to their Arms.

About the end of Four and twenty hours, another Company mounted the Guard, and relieved these. In the mean time, whilst I waited with a great deal of Melancholy, to know what it would please Fortune to determine of me; amidst my Misfortunes my charitable Magpy informed me of all that past.

Amongst other things, I remember she told me, That the Mobile of the Birds, had made a heavy Rout, because they had kept me so long without devouring me; that they had alledged I would grow so lean, that there would be nothing of me left but Bones to pick.

The

The Clamor was like to break out into a Sedition; for my Magpy having taken the boldness to represent, That it was a barbarous Procedure to put to death in that manner without any Tryal, an Animal, which in some sort had Reason as well as they; they had like to have torn her to pieces, pretending that it would be very ridiculous, to think that an Animal stark naked, which Nature her self, when she produced it, took no care to furnish with things necessary for its Preservation, should be capable of Reason like them: Nay more, added they, were it an Animal that resembled our shape somewhat more, it were somewhat; but of all things the most unlike, and most hideous. In a word, a bald Beast, a featherless Bird, a Chimera patched up of all sorts of Natures, and terrible to all. Man, I say, so vain and foolish, that he perswades himself, we were only made for him. Man, who with his sharp-sighted Soul, cannot distinguish Sugar from Arsenick, and who'll swallow down Hemlock, when his quaint Judgment hath made him take it for Parsley. Man, who maintains that there is no Reasoning, but with a reference to the Senses; and yet has the weakest, dullest and most erring Senses of all Creatures living. Man, in fine, whom Nature hath

formed

formed as she doth Monsters, to shew her skill; and yet hath filled with the Ambition of domineering over, and extirpating all other Animals.

This was the discourse that the wiser sort held; as for the Rabble they cried, That it was a horrid thing, to believe that a Beast who had not a Face like theirs, should have Reason. Now, chattered they one to another, he hath neither Beak, Feathers nor Claws, and can he have a Spiritual Soul? Strange! what Impertinence is that!

Though the more generous pittyed me, yet that hindered not, but that an Inditement was brought in against me; all the Writings were drawn upon the Bark of a Cypress Tree: And then some Days after, I was carried before the Tribunal of Birds. There were none chosen for Advocates, Counsellors and Judges of the Court, but Magpys, Jays, and Starlings; because they would have none but those that understood my Language. Instead of trying me at the Bar, they set me stradling upon a Stump of a rotten Tree; where he that was President of the Court, having chattered two or three times with his Beak, and Majestically shrugged his Feathers, asked me, From whence I came, of what nation and kind I was? My charitable Magpy had given

me

me some Instructions before, which were very useful to me; and amongst other, that I should have a special care, not to confess that I was a Man. I made answer then, That I was of that little World which is called the Earth, that the Phenix, and some others whom I saw in Court, might have told them of: That the Climat where I was born, lay under the temperate Zone of the Artick Pole, in the utmost part of *Europe*, which was called *France*; and that as to my kind, I was not a Man, as they imagined, but an Ape; That some Men had carried me away very young out of the Cradle, and brought me up amongst them: That their bad Breeding had rendered my skin so delicate; that they made me forget my Mother's Tongue, and taught me theirs: That to comply with these Wild Animals, I had accustomed my self to walk only on two Legs: And that, in a word, since it is more easie to degenerate than to improve in kind, the Opinion, Custom, and Food of these unclean Beasts, had so much power upon me, that hardly my Parents, who are Apes of Quality, could themselves know me. I added for my Justification, that they might have me viewed by expert Birds, and that in case they found me to be a Man, I should be willing to be annihilated as a Monster. Gen-

Gentlemen, cried a Swallow of the Court, so soon as I had made an end, I hold him as convicted: You have not forgot that just now he said, That the Country where he was born, was *France*; but you know that the Apes in *France* do not engender: Judge now after that, if he be what he boasts himself to be.

I made answer to my Accuser, That I was carried away so young out of the Bosom of my Parents, and Transported into *France*, that I had reason to call that my native Country which I remembred best.

That reason though specious, was not sufficient; but the most part being ravished to hear, that I was not a Man, were very willing to believe it so: For those who had never seen any, could not be perswaded, but that a Man was a far more horrid Creature, than I appeared to be to them; and the discreeter added, That a Man was so abominable a thing, that it were good they should believe him to be only an imaginary Being.

All the Court clapt their Wings for Joy, and immediatly I was commited to the Aldermen-Birds to be examined, on condition they should present me again next Morning, and at the opening of the Court, make report of the Verdict of the Jury. They undertook the Affair then, and carried me

into

into a Bye-copice: Whilst they kept me there, they did nothing but play a hundred Monkey tricks about me; sometimes they knocked their Feet one against another, by and by they dug little Holes, only that they might fill them up again; and then all of a sudden I was amazed that I could see no Body at all.

The Day and Night were spent in these trifles, until next Morning, that the hour appointed being come, they brought me to appear again before my Judges: Where my Syndicks being desired to tell the truth, answered, That to discharge their Consciences, they thought themselves obliged to inform the Court, That for certain I was not an Ape, as I bragg'd: For, said they, it was to no purpose for us to leap, skip, whirl round, and invent before him a hundred odd Tricks and Capers, whereby we thought to move him to do the like, according to the custom of Apes, when all would not do. Now though he hath been bred amongst Men, seeing an Ape is always an Ape, we maintain that it could not have been in his power, to refrain from imitating our Apish Tricks. This, Gentlemen, is our Report.

The Judges at that drew near to give their Opinions; but they perceived that the Sky was overcast and charged with

Clouds,

Clouds, which made them break up the Court.

I imagined that the appearance of bad Weather inclined them to do so.

When the Attorney-General by order of Court came, and gave me intimation, That they would not pass Sentence on me that Day; that they never determined a criminal Process, when the Sky was not serene; because they were afraid, that the bad temper of the Air, might work some alteration in the good Constitution of the minds of the Judges; that the peevish temper the Birds are in, during Rain, might influence the Cause: Or, in a word, that the Court might revenge their Sadness upon the Prisoner; and therefore it was, that my Sentence was deferr'd till fairer Weather. I was carried back to Prison then, and I remember that in the way my charitable Magpy never forsook me; she kept flying all along by my side, and I believe she would never have left me, had not her Companions drawn near to us.

At length I arrived at the place of my Prison, where, during, my Captivity, I had no other Food but the King's Bread; for so they call half a hundred Worms, and as many Maggots, that every seven hours they brought me to eat.

I thought that I should have appeared again next day, and every Body thought so too; but one of my Keepers told me, at the end of five or six Days, that all that time had been employed, in rendring Justice to a Society of Gold-finches, who had complained of one of their number. I asked my Keeper, What crime that poor Wretch had been accused of? Of the most enormous, replied my Keeper, that a Bird can be aspersed with. They accuse it------ can you believe it? They accuse it------ but good Gods! the very thoughts of it makes my Feathers to stand an end! In a word, they accuse it, that during the space of Six Years, it hath not as yet deserved to have a Friend; and therefore it hath been condemned to be a King, and a King of a People that differ from it in kind.

Had its Subjects been of its own nature, it might at least have beguiled its Eyes and Desire with their Pleasures: But seeing the pleasures of one kind, have no relation to those of another, it will support all the fatigues, and tast all the bitterness of Royalty, and never be able to relish the pleasures thereof in the least.

They have sent it away this Morning, accompanied with a great many Physitians, to take heed that it do not poison it self by the way. Though

Though my Keeper was naturally a great Talker, yet he durst not entertain me any longer in discourse, for fear of being suspected of Intelligence with me.

About the end of the Week, I was again brought before my Judges.

They rested me upon the breech of a little Tree without Leaves.

All the Birds of the Long-Robe, as well Advocates, Counsellors, as Judges, and Presidents, roosted by Stories, every one according to his Dignity, on the Top of a tall Cedar: For the rest who were only present out of Curiosity, they placed themselves promiscuously till all the Seats were full, that's to say, till the Branches of the Cedar were covered with Feet.

The Magpy, in whom I observed all along so much Compassion for me, came and perched upon my Tree, where pretending to divert her self by pecking the Moss: Really, said she to me, you cannot believe how much I am concerned at your Misfortune; for though I am not ignorant that amongst the Living, a Man is a Plague, that ought to be purged out of all well-govern'd States; yet when I call to mind, that I was bred amongst them from the Cradle, that I have learned their Language so perfectly, that I had almost forgot mine own; and that I have eaten out of their Hands

such

such excellent Green Cheese, I cannot think on't, but that it brings Water into my Eyes and Mouth; I have so great a kindness for you, that I cannot incline to the right side.

She had gone on so far, when we were interrupted by the coming of an Eagle, that lighted amongst the Branches of a Tree, pretty near to mine. I was about to have risen and fallen upon my knees before the Eagle, thinking he had been the King, if my Magpy with her Foot had not held me fast in my Seat. Did you think, said she, that that great Eagle had been our Sovereign? That's an Imagination of you Men, who because you suffer your selves to be commanded by the greatest, the strongest, and the most cruel of your Companions, have foolishly thought, judging of all things according to your own measures, that the Eagle ought to command us.

But our Politicks are quite different, for we never chuse for our Kings, but the Weakest, the Wildest, and most Peaceable: Nay, and we change them every Six Months, and pitch upon the Weak, to the end, that the meanest amongst us, who may have been wronged by him, may take his Revenge. We chuse the Mild, to the end he neither hate, or be hated of any

any Body; and we would have him to be of a Peaceful Temper, for avoiding of War, the Sink of all Injustice.

Once every Week he holds a Parliament, where all are received to propose their Grievances against him. If there be but three Birds only dissatisfied with his Government, out he goes; and they proceed to a new Election.

All that Day the Parliament sits, our King is mounted on the top of a high Yew-Tree, upon the brink of a Lake, bound Feet and Wings. All the Birds one after another pass before him; and if any of them know him to be guilty of a Crime that deserves death, he may throw him into the Water; but he must upon the spot, justifie the fact by good Reasons, otherwise he is Condemned to the said Death.

I could not forbear to interrupt, and ask her, what she meant by the said Death? And this is the Answer she made me:

When the Crime of a Malefactor is judged to be so enormous, that an ordinary Death is not sufficient to expiate it, they endeavour to chuse one that contains the pain of many; and in this manner they proceed to it:

Those amongst us that have the most melancholick and doleful Tone, are sent
to

to the Malefactor, who is carried upon a dismal Cypress. There these sad Musicians gather about him, and by the Ear fill his Soul with such tragical and doleful Notes, that the bitterness of his Sorrow disordering the Oeconomy of his Organs, and pressing his Heart, he pines away to the sight, and dies choaked with Sadness.

However such a spectacle never happens; for seeing our Kings are exceeding mild, they never force any Body to incur so cruel a Death, upon the account of Revenge.

He that at present Reigns is a Dove, who is of so peaceable a temper, that t'other day, when two Sparrows were to be made Friends, it was the hardest thing in the World, to make him conceive what Enmity was.

My Magpy could not continue so long a discourse, without being observed by some of the By-standers; and because she was already suspected of some Intelligence with me, the chief of the Assembly made one of the Eagles of my guard, catch her by the Neck, and make sure of her Person. King Dove arrived in the mean while; all were silent, and the first thing that was heard, was the complaint of the great Censor of the Birds, which he made against the Magpy. The King being

ing fully informed of the Scandal she had given, asked her her Name, and how she came to know me? Sir, answered she, all in amaze, My name is *Magget*, there are here a great many Birds of Quality, that will vouch for me. One day in the World of the Earth, of which I am a Native, I was informed by *Chirpper* the Posy there, (who having heard me cry in my Cage, came to visit me at the Window where I hung,) that my Father was *Bobb-tail*, and my Mother *Crack-nuts*: I had not known so much but for him; for I was carried away very Young, from under the Wings of my Parents; my Mother sometime after died of Grief; and my Father, being then past the Age of having any more Children, despairing to see himself without Heirs, went to the War of the Jays, where he was killed by a peck in the brain. They that carried me away were certain wild Animals, whom they call Hog-herds; who had me to be sold at a Castle, where I saw that Man who now stands upon his Tryal. I cannot tell, whether he conceived any Kindness for me, but he took the pains to cause the Servants to cut meat for me; he had sometimes the goodness to prepare it for me himself. If I catcht cold in the Winter, he carried me to the Fire, lined my Cage, or ordered the Gardiner to warm me in his Bosom.

The

The Servants durst not vex me in his presence: and one day I remember, he saved me from the Jaws of the Cat, who held me in her Paws, to which my Lady's little Page had exposed me; but it will not be impertinent, to tell you the Cause of that Barbarity. To comply with *Verdelet* (for that was the Page's name,) I was chattering one day some idle words that he had taught me. Now it happened, as ill Luck would have it, though I always repeated my Lessons in course, that I came to say in order, just as he came in to deliver a Message, *Be quiet, you Son of a Whore, you lye.* The Man there that stands Indicted, who knowing the Rogue to be naturally given to Lying, imagined, that I might very well have spoken by Prophecy, and sent to the place to know, if *Verdelet* had been there: *Verdelet* was convinced of his Knavery, *Verdelet* was whipt, and *Verdelet* in revenge, would have had me devoured by *Maulkin*. The King by a Nod of the head shew'd, that he was satisfied with the Pity that she had conceived for my disaster: However he discharged her to speak any more to me in private. Then he asked my Adversary's Council, if his Plea was ready? He made a sign with his Foot, that he was going to speak, and if I mistake it not, these are the Points whereon he infisted againſt me:

The Plea brought in the Parliament of Birds, Aſſembled againſt an Animal, accuſed of being a Man.

Gentlemen, The Plaintiff is *Guillemot* the Fleſhy, a Partridge by extraction, lately arrived from the World of the Earth, his Breaſt ſtill gaping by a ſhot that he hath received from Men; Demandant againſt all Mankind, and by conſequence againſt an Animal, whom I pretend to be a member of that great Body. It would be no hard matter for us, to hinder, by his death, the Violences that he can commit: Nevertheleſs, ſeeing the Preſervation or Loſs of every thing that has breath, concerns the Common-Wealth of the Living, I think we ſhould deſerve to have been made Men, that's to ſay, Degraded from Reaſon and Immortality, which we enjoy above them, had we reſembled them in any unjuſt Action like theirs.

Let us examine then, Gentlemen, the Difficulties of this Cauſe, with all the Application that our divine minds are capable of.

The ſtreſs of the matter lies here, to wit, Whether or not this Animal be a Man? and

and then in cafe we make it out, that he is, whether or not he deferves Death for that?

For my part, I make no doubt but that he is; in the firft place, Becaufe he is fo impudent as to tell a Lye, in maintaining that he is not; in the fecond place, Becaufe he laughs like a Fool; thirdly, In that he weeps like a Sot; fourthly, In that he blows his Nofe like a nafty Villain; fifthly, In that he is Feathered but in part; fixthly, In that he carries his Tail before; feventhly, In that he hath always a great many little fquare Bones in his Mouth, which he has neither the wit to fpit out, nor fwallow down; eighthly and laftly, Becaufe every Morning he lifts up his Eyes, his Nofe, and large Snout, claps his open Hands clofe together, which he points up to Heaven, joins them into one piece, as if it troubled him to have two at liberty; breaks his Legs fhort off by the middle, fo that he falls upon his Geegots; and then by Magical words that he mutters, I have obferved, That his broken Legs are knit again, and that he rifes up as gay as he was before. Now, you know, Gentlemen, that amongft all Animals, none but Man has fo black a Soul, as to be given to Magick, and by confequence I conclude, That this is a Man: We are

now

now to enquire, whether or not, as Man, he deserves to be put to death.

I think, Gentlemen, it never was yet doubted, but that all Creatures are produced by our common Mother, to live together in Society. Now if I prove, that Man seems to be Born only to break it; shall I not make it out, that he going contrary to the end of his Creation, deserves that Nature should repent her self of her work?

The first and fundamental Law, for the maintenance of a Republick, is Equality: But Man cannot endure it to Eternity; he falls upon us that he may eat us; he perswades himself that we were only made for his use; he makes the Barbarity, wherewith he massacres us, and the small Resistance he finds on our side, an Argument of his pretended Superiority: And nevertheless wo'nt own Eagles, Condores, and Griffins, who are too hard for the strongest Man, to be his Masters.

But why should that great Size, and conformation of Members, make the diversity of Kind, seeing there are Dwarffs and Giants to be found amongst Men themselves?

Nay more, that Empire wherewith they flatter themselves, is but an imaginary Right: On the contrary, they are so inclinable

nable to Servitude, that leaſt they ſhould not ſerve, they ſell one another for Slaves. In this manner, the Young are Slaves to the Old, the Poor to the Rich, the Clowns to the Gentlemen, the Princes to the Monarchs, and the Monarchs themſelves to the Laws which they have Eſtabliſhed. And beſides all that, the poor Drudges are ſo afraid to be without Maſters, that as if they apprehended, that Liberty might come to them from ſome unexpected place; they frame to themſelves Gods in all parts, in the Water, in the Air, in the Fire, and under the Earth; they'll make them of Wood, rather than want; nay, I fancy alſo that they tickle themſelves with the vain hopes of Immortality, not ſo much out of a Horrour, that they have of being annihilated, as for fear that they may have none to command them after their death. Here's the fine effect of that fantaſtical Monarchy, and of that natural Empire of Man, as they would have it, over the Animals, nay and over us too; for he has been ſo inſolent, as even to pretend to that. In the mean while, in conſequence of that ridiculous Principality, he fairly takes to himſelf the power of Life and Death over us; he lays ſnares for us, chains us, claps us up in Priſon, kills us, eats us; and makes the power of killing thoſe which remain free, a mark of Nobility: He thinks that the Sun is lighted on purpose to let him see how to make War against us; that Nature hath only suffered us to take our turns in the Air, that from our flight he may draw lucky or unlucky Auspices; and that when God put Entrails into our Bodies, his intention only was, to make a great Book, wherein Man might learn the Science of future Contingencies.

Good, then, is not this unsupportable Pride? Could any that's guilty of it, deserve a less punishment, than to be made a Man? However, I don't insist upon this as a reason, why ye should condemn this Man: The poor Beast wanting the use of Reason that we have, I excuse those errors of his, that proceed from want of Judgment; but for such as are only the Daughters of his Will, I demand Justice. For instance, in that he kills us, though we do not attack him; in that he eats us, when he may satisfie his hunger with more convenient Food; and what I esteem the basest of all, in that he debauches the good nature of some of our Brethren, as of Lanners, Faulcons and Vultures, by teaching them to murder those of their kind, and to feed on their fellow Creatures alive; or to deliver us up into his clutches.

That

That alone is so pressing a Consideration, that I beg the Court he may be dispatched by the sad Death.

The whole Bench shivered for horror at so terrible a Punishment; and therefore that they might have ground to moderate it, the King made a Sign to the Council, that was assigned me, to answer.

This was a Starling, and a great Lawyer, who having three times stampt with his Foot, upon the Branch he sat on, spake to the Court in this manner:

It is true, Gentlemen, that moved with Pity, I undertook the defence of that unfortunate Beast; but just as I was about to Plead, I felt a remorse of Conscience, and, as it were, a secret Voice, that hath forbidden me to fulfil so detestable a Resolution: So that, Gentlemen, I declare to you, and the whole Court, That for the Salvation of my Soul, I'll not contribute in any manner, to the preservation of such a Monster, as Man is.

The whole Mobile clacked with the Beak, in sign of Joy, and to congratulate the Sincerity of so Conscientious a Bird.

My Magpy offered to Plead for me in place of the other, for it was impossible for her to be heard; because that being bred amongst Men, and perhaps, infected with their

their Morality; it was to be feared, that she would manage the Cause, with a prejudicated Mind; for the Court of Birds, never suffer a Lawyer, that concerns himself more for the one Client, than for the other, to be heard; unless he make it appear, That that Inclination proceeds from the Parties being in the Right.

When my Judges saw, that no Body appeared in my defence, they stretched out and shook their Wings, and immediately flew to Voting.

The greatest part, as I was informed since, insisted hard, that I should be dispatched by the sad Death; but nevertheless, when they perceived that the King inclined to Clemency, all joined with him in Opinion. Thus my Judges moderated themselves, and instead of the sad Death, which they excused me from, they thought it convenient, that my Punishment might quadrate with some of my Crimes, and I destroyed by a Death, which might serve to undeceive me, of that pretended Empire of Man, over the Birds, which I bragg'd of, that I should be abandoned to the weakest of those that are carried by Wings; my meaning is, That they Condemned me to be eaten up by Flies.

At the same time the Court broke up, and I heard a whisper, that they had not
enlarged

enlarged in Specifying the particular Circumstances of my Tragedy, becaufe of an accident that happened to a Bird, who juft as he was about to fpeak to the King, had fallen into a fwoon. It was thought to have been occafioned, by the Horrour that had feized him, in looking too ftedfaftly upon a Man: And therefore I was ordered to be carried away.

But my Sentence was pronounced firft; and as foon as the Ofpray, which officiated as Clerk to the Affifes, had made an end of reading it to me, I perceived all about the Sky blackened with Flies, Drones, Bees, Gnats and Muskettoes, which hummed for impatience.

I expected that my Eagles fhould have come and carried me away after the ufual manner; but in place of them a great black Oftridge came, that ignominioufly fet me ftradling upon his back; for amongft them, that's the moft difgraceful pofture a Malefactor can be put into; and no Bird for what offence foever, can be Condemned to it.

The Officers that waited on me to Execution, were half a hundred Condores, and as many Griffins in the van; after whom came flying foftly, a proceffion of Ravens, that croaked I know not what mournful Ditty; and I fancy that I heard as at a greater diſtance, a Chorus of Owls, that anſwered them.

As we parted from the place where I had received Sentence, Two Birds of Paradice, who had orders to aſſiſt me at my Death, came and ſat on my Shoulders.

Though my Soul, at that time, was very heavy and diſcompoſed, by reaſon of the lamentable Condition I was in; yet I remember, in a manner, all the Reaſons they made uſe of to comfort me.

Death, ſaid they to me, (putting their Beak to my Ear,) without doubt is no great Evil, ſeeing Nature our good Mother ſubjects all her Children unto it; and it ought not to be a matter of great Conſequence, ſince it happens at all times, and for the leaſt things: For if Life were ſo excellent, it would not be in our power, not to give it; or if Death were attended by conſequences of Importance, as thou imagineſt, it would not be in our power to give it: There is a great deal of appearance to the contrary, ſeeing the Animal begins by play, and ends at the ſame rate. I ſpeak to thee in this manner, becauſe thy Soul not being immortal as ours is, thou mayeſt very well conclude, That when thou dieſt, all dies with thee. Let it not trouble thee then, that thou doeſt a little ſooner, what ſome of thy fellows

fellows will do e're it be long: Their condition is more deplorable than thine; for if Death be an evil, it is only Evil to those who are to die: And in respect of thee, who hast not above an hour betwixt here and there; they shall be Fifty or Sixty Years in a state of dying; and besides, mind me, he that is not born is not unhappy. Now thou art going to be like him that is not born: In a twinkling of an Eye, when thou art out of this Life, thou shalt be what thou wast a twinkling of an Eye before; and that twinkling of an Eye being over, thou shalt be as long dead, as he that died a Thousand Ages ago: But make the worst on't; suppose Life be a Blessing, the same Accident, that in the infinite spaces of time hath made thee to be; may it not some time or other, make thee once more to be again? Matter, which by various mixtures, arrived at length to that Number, Disposition and Order, necessary for the Construction of thy Being; may it not, by mixing again of new, attain to a Disposition requisite for bringing thee once more again into Being? Yes, it may; but thou'lt say to me, I shall not remember that I have been. Ha! dear Brother, what does that concern thee, provided thou findest thy self in Being? And then may it not be, that to comfort thee for the loss of thy Life, thou'lt
ima-

imagine the same Reasons which I at present propose to thee?

These considerations are weighty enough to make thee patiently drink that bitter Potion; yet I have others more pressing still, which will without doubt incline thee to wish for it. Thou must be perswaded, Brother, that as thou and the rest of Brutes are material, and that as Death, instead of annihilating Matter, does only trouble the Oeconomy thereof; so thou oughtest, I say, certainly to believe, that ceasing to be what now thou art, thou'lt begin to be something else. Grant then, that thou only become a clod of Earth, or a Pibble, thou'lt still be somewhat less wicked than Man. But I have a Secret to discover to thee, which I would not any of my Companions should hear from my Mouth, and that is, That being Eaten, as thou art going to be, by our little Birds, thou'lt pass into their Substance; yes, thou'lt have the honour to contribute, though blindly, to the Intellectual operations of our Flyes, and share of this Glory, that if thou reasonest not thy self, thou'lt make them at least to reason.

About this part of the Exhortation, we arrived at the place appointed for my Execution.

There

… *World of the Sun.*

There were Four Trees very near, and in a manner equally diftant one from another, on every one of which at a paralel height, a great Heron perched. They took me down from the Black Oftridge, and a great many Cormorants lifted me up to the place where the Herons were. Thefe Birds being oppofite to one another, and firmly perched on their feveral Trees, with their prodigious long Necks, as with a Cord, twifted about me, the one about my Arms, the other about my Legs, and bound me fo faft, that though every one of thefe members was only tied by a fingle Neck, yet it was not in my power, to wag or ftir in the leaft.

They were to continue a long while in that pofture; for I heard orders given to thofe Cormorants that lifted me up, to go and fifh for the Herons, and to flide their Foot into their Beaks.

The Flies were ftill expected, becaufe they could not fly fo faft as we had done; however it was not very long before they were heard.

The firft thing they exploited, was to diftribute my Body among them into feveral Provinces; and that Diftribution was fo malicioufly made, that my Eyes were affigned to the Bees, to the end they might Sting them out as they fed on them;

my Ears to the Beetles and Drones, that they might ftum and devour them at the fame time; my Shoulders to the Musket-toes, to the end, they might make me itch with their Bitings; and fo of the reft. No fooner had I heard them fettle their Orders, but that immediately I faw them approach. All the Atoms and Motes in the Sun, feemed to be converted into Flies; for fcarcely was I vifited with two or three faint Beams of Light, that feemed by ftealth to reach me, fo clofe were thefe Battalions, and fo near my Flefh.

But as every one of them was greedily chufing the place where he was to bite, I perceived them to recoyle briskly all of a fudden; and amidft the confufion of an infinite number of Shouts, that made the Clouds refound again; I feveral times diftinguifhed the word Pardon, Pardon, Pardon.

Afterward Two Turtle Doves drew near to me; at their approach, all the ghaftly Preparatives for my Death were diffipated: I felt my Herons let loofe the twifts of their long Necks, wherewith I was begirt, and my Body extended in form of a St. *Andrew's*-crofs, flide from the Top of the Four Trees, down to their very Roots.

I

I expected no less from my fall, than to be bruised by some Stone in the Ground; but when my fear was over, I was not a little surprised, to find my self decently seated upon a White Ostridge, who fell a galloping, so soon as he felt me upon his Back.

They made me take another way, than the way I came; for I remember that I crossed a great Wood of Myrtles, and another of Turpentine Trees, adjoining to a vast Forrest of Olive Trees, where King Dove in the middle of his Court staid for me.

So soon as he perceived me, he gave a sign that they should help me to get down. Immediately Two Eagles of the Guard, gave me their Talons, and carried me to their Prince.

I would in honour, have embraced and kissed his Majesty's little Claws, but he drew back: And I ask you the question, said he first, If you know that Bird?

At these words they shewed me a Parrot, who began to turn round and clap his Wings, when he perceived that I considered him: Yes, I fancy, cried I to the King, that I have seen him somewhere; but Fear and Joy have so confounded my Memory, that I cannot as yet call distinctly to mind, where it has been.

With that the Parrot came, and embracing my Face with its Wings, said to me, How! do'nt you know *Cæsar*, then, your Cousins Parrot, which hath so often given you the occasion, to maintain that Birds used Reason? It is I who just now had a mind, after your Tryal was over, to declare the obligations I have to you; but grief to see you in so great danger, made me fall into a Swoon. His discourse fully opened my Eyes, and having known him perfectly, I embraced and kissed him, and he embraced and kissed me. Is it thee, then said I, my poor *Cæsar*, whose Cage I opened to give thee thy Liberty, that the Tyrannical Custom of our World deprived thee of?

The King interrupted our Caresses, and spoke to me in this manner: Man, with us a good Action is never lost; and therefore it is that, tho as being a Man, thou deservest to die, only because thou wast Born, yet the Senate gives thee thy Life. It's fit we should thus acknowledge those Notices, wherewith Nature enlightened thine Instinct, when it gave thee a Foretast of that Reason in us, which thou wast not capable of understanding. Go then in Peace, and live Joyfully.

He gave some orders in private, and my White Ostridge, conducted by the Two

Turtle Doves, carried me out of the Assembly.

My Bird having galloped with me about twelve hours, left me near to a Forest, into which I went so soon as it was gone. There I began to tast the Pleasure of Liberty, and of feeding on the Honey, which distilled down the Bark of the Trees.

I fancy that I should never have made an end of my Walk, (for the agreable Varieties of the place, presented always some new thing more delightful to my Eyes,) had my Body been able to hold out: But seeing I found my self at length wholly overcome with Wearisomness, I softly laid my self down upon the Grass.

Lying thus stretched out under the shadow of the Trees, I found my self invited to Sleep, by the fresh Air and Solitude of the Place; when a humming noise of confused Voices, which seemed to sound about my Ears, made me awake with a start.

The ground appeared to be very level and smooth, without the least Bush that might intercept the Sight; and therefore my Prospect reached far amongst the Trees of the Forest. Nevertheless, the Murmuring that I heard, could not be but very near me; so that listning to it more attentively, I very distinctly heard a parcel

cel of *Greek* Words; and amongst several that discoursed together, one who spake to this purpose,

Doctor, one of my Allies, *Elm* with Three Heads, hath just now sent me a Chaffinch, to acquaint me that he is Sick of an Hectick Feaver, and of a Scurffy Moss, wherewith he is covered from Head to Foot. I beg of you for all Love, that you would be pleased to order him somewhat.

For some time I heard no more; but after a little intermission, I thought I heard one make this reply: Though *Elm* with Three Heads, were none of your Allies; and though instead of you who are my Friend, the greatest Stranger of all our kind, should desire the same thing of me, I would grant it; for my Profession obliges me to assist every Body. You shall therefore acquaint *Elm* with Three Heads, That for the Cure of his Distemper, it is necessary, that he Suck in as much humide, and as little dry Nourishment, as possibly he can; that for that end he ought to direct the little Fibres of his Root, towards the moistest place of his Bed: That he must keep himself Merry and Chearful, and daily be diverted by a consort of Musick, of some excellent Nightingals. You'll hear from him afterwards, how he finds him-

himself, with that course of living; and then according to the Progress of his Distemper, when we have prepared his Humours, some Stork of his Friends shall from me give him a Glister, that will fully recover him.

These words being ended, I heard not the least noise more; till about a quarter of an hour after, that a voice which I fancy I had not observed before, came to my Ears, and spake in this manner, *Hola, Gaffer Forked Trunk*, what, do ye Sleep? I heard another voice that thus replied, No, *Fresh-bark*, why? Because, said the first that spake, I find the same Emotion in me, that commonly we do, when these Animals they call Men, come near us; and I would ask you, if you feel the same thing?

It was some time before the other made Answer; as if he intended to employ the most exquisite of his Senses in that Discovery; but at length he cried out: Good God, you are in the right, and I swear to you I find my Organs so full of the Ideas of a Man, that I am the most mistaken in the world, if there be not one very near to this place: At that time there was a mixt noise of voices that said, they smelt out a Man.

It was in vain for me to look about on all hands; I could not difcover whence that word could come: At length being a little recovered from the Horror, whereinto that accident had caſt me; I made anſwer to the voice, which I took to be that which had asked, if there was any Man there; that there was one: But I befeech you, continued I, whoever you be that fpeak to me, tell we who you are. Within a trice after I heard thefe words:

We ſtand in thy Prefence; thine Eyes behold us; and yet thou feeſt us not: Look upon thefe Oaks, on which we perceive thine Eyes are fixt: They are we that fpeak to thee; and if thou be furprized that we fpeak a Language, ufed in the World from whence thou comeſt; know, that our firſt Parents were Natives of it: They lived in *Epirus*, in the Foreſt of *Dodona*; where their natural goodnefs inclined them to render Oracles to the afflicted, who confulted them. For that end they learnt the *Greek* Tongue, which at that time was moſt univerfal; that fo they might be underſtood: And becauſe we are defcended of them, from Father to Son, the gift of Prophecy hath been tranfmitted even to us. Now thou muſt know, That a great Eagle, to whom our Fathers of *Dodona* had given retreat; being difabled from going

to

to prey, becaufe fhe had broken one of her Leggs, fed upon the Acorns, which their branches furnifhed her with: When one day, being impatient of living in a World where fhe fuffered fo much, fhe took a flight to the Sun; and profecuted her Voyage fo happily, that at length fhe arrived in the luminous Globe, where we are: But upon her arrival, the heat of the Climate put her into a reaching to Vomit, which made her bring up a great quantity of Acorns, not as yet digefted; thefe Acorns fprouted, and produced the Oaks which were our Progenitors.

In this manner we changed our Habitation: Neverthelefs, though you hear us fpeak a humane Language, you muft not therefore conclude, That the other Trees exprefs themfelves fo: No Trees, but we Oaks defcended from the Foreft of *Dodona*, fpeak as you do. For as for the other Vegetables, they exprefs themfelves after this manner. Have you not minded that foft and gentle Breeze, that never fails to blow about the Skirts of a Wood? That's the breath of their Words; and that eafie Murmuring, or delicate Whifper, whereby they break the facred filence of their folitude, is, to fpeak properly, their Language. But though the noife of Forefts feem always to be the fame; yet it is fo different, that every kind of Vegetables have distinctly their own; so that the Birch speaks not like the Maple, nor the Beach like the Cherry Tree: Had the Foolish People of your World, heard me speak as I do, they would have thought it had been a Devil, enclosed within my Bark; for they are so far from believing that we can reason, that they do not imagine we have a Sensitive Soul; though every day they see, that at the first blow the Woodman gives the Tree, the Hatchet enters the Wood four times deeper, than at the second; and they ought to conjecture, that the first blow surprised it, and took it unprovided; since that as soon as it is warned by the pain, it contracts within it self, unites it Forces for resisting, and in a manner petrified, that it may withstand the sharpness of its Enemies Arms. But it is not my design, to make blind Men Judges of Colours; an Individual is to me the whole kind, and the whole kind, is no more to me than an Individual, when the Individual is not infected with the Errors of the kind; and therefore be attentive, for in speaking to you, it's the same thing to me, as if I spoke to all Mankind.

You are to understand then, in the first place, That almost all the warbling Consorts of the Musick of Birds, are composed

in

in praife of Trees; but in recompence alfo of the pains they take, in celebrating our famous Actions, it is we that take care to hide their Amours; for don't you imagine when it cofts you fo much trouble to find one of their Nefts, that that's occafioned by the Sagacity wherewith they hide it? No, it is the Tree it felf, that hath twifted its Boughs about the neft, to fecure the Family of his Lodger, from the Cruelties of Man: And on the contrary, confider the Airies of thofe, which are hatched either for the deftruction of Birds, their fellow Citizens, fuch as Sparrow-Hawks, Hobbies, Kites, Faulcons, &c. or which only fpeak to breed Quarrels, as Jays and Magpies; or that delight to frighten us, as Owls and Howlets: You fhall obferve, that the Nefts of fuch are expofed to the fight of all People; becaufe the Tree removes its Branches from them, that it may leave them for a Prey.

But there is no need of fpecifying fo many things, to prove that Trees exert your Functions, as well in Mind as in Body. Is there any one amongft you, who hath not obferved, that in the Spring, when the Sun hath refrefhed our Bark with a fertile Sap, we thruft out our Branches, and extend them loaded with Fruit, upon the Breafts of the Earth, that we are in Love withal?

withal? The Earth on her side opens, and is warmed with the like heat; and makes her approaches towards a Conjunction, whilst our Branches discharge into her Lap, that which she so ardently desires to conceive. She is, however, Nine Months in breeding and forming that *Embrio*, before she bring it forth; but the Tree her Husband, fearing that the Winters cold may be prejudicial to her Conception, strips himself of his green Garment to cover her, and contents himself with an old *Fuil-demort* Cloak, to hide part of his Nakedness.

Well then, O Men, you look eternally on these things, and never see them: Nay, more convincing Proofs have presented themselves to your Eyes; but none are so Blind, as those that will not See.

I listened most attentively to the discourse, with which that Arboreal Voice entertained me, and was expecting the sequel; when all of a sudden it broke off with a Tone like to the Whizzing of the shortness of Breath, that hinders one to speak.

When I perceived it obstinately resolved to be silent, I adjured it by all, which I thought might most affect it, that it would vouchsafe to instruct one, who had run the Risk of so long and dangerous a Voyage, upon the account only of learning.

At

At the same time I heard Two or Three Voices, which for my sake made the same request to it, and one I distinguished that said to it, as if in anger,

Well then, since you complain so much of your Lungs, repose your self; I'll tell him the Story of the Amorous Trees.

Whoever you be, cried I, falling upon my Knees, O Wisest of all the Oaks of *Dodona*, who condescendest to take the pains to instruct me, know this; That you shall not teach an ungrateful Person; for I vow, that if ever I return to my native Globe, I shall publish the Wonders, you are pleased to make me a Witness of. I had no sooner made this Protestation, but I heard the same voice proceed in this manner: Look, Little Man, and you shall see about Fourteen or Fifteen steps to the Right Hand, Two Twin-Trees of a middle Stature, which confounding their Branches and Roots, strive by all possible means to unite and become but one.

I turned my Eyes towards these Plants of Love, and observed that the leaves of both gently stirred, as it were, by a voluntary Motion, excited by their Agitation so delicate a murmur, that hardly it grazed upon the Ear; and yet one would have said, that thereby they mutually asked, and answered one another.

Having

Having spent as much time as was necessary, to observe that double Vegetable, my good Friend the Oak went on in his discourse, after this manner:

You cannot have lived to this Age, and not have heard of the celebrated Friendship of *Pylades* and *Orestes*.

I would describe to you all the Joys of a sweet Passion, and tell you the Wonders wherewith these Lovers astonished their Age, did I not fear that so much Light might offend the Eyes of your Reason; and therefore I shall paint those two young Suns only in their Eclipse.

Let this then suffice you to know, That one day the brave *Orestes* being engaged in a Battle, sought out for his dear *Pylades*, that he might have the Pleasure of overcoming or dying in his Presence. When he perceived him amidst an hundred Arms of Iron, lifted up over his Head: Alas! what became of him? In despair he threw himself through a Forest of Pikes: He cried, roared and foamed: But how ill do I express the fearful Commotions of that Inconsolable Man; he tore his hair, bit his Hands, rent his Wounds; nay, and when I have said all I can say, I am obliged to confess, that the means of expressing his grief, died with himself. When he thought to cut out a way with his Sword to get to the

the affiftance of *Pylades*, a Mountain of Men withftood his paffage. Neverthelefs he broke through them; and having long marched upon the Bloody Trophies of his Victory, by little and little he approached to *Pylades*; but *Pylades* feemed to him already fo near Death, that he durft hardly refift his Enemies any longer, for fear he might furvive the thing for which alone he lived. To fee his Eyes already full of the fhades of Death, one would have faid, That he endeavoured by his Looks to poyfon the Murderers of his Friend. At length *Pylades* fell down dead; and amorous *Oreftes* perceiving his own life, to be upon the brink of his Lips, ftill retained it; till with a wandring look, having fought and found out *Pylades* amongft the Dead, he feemed, by joining Mouth to Mouth, as if he intended to infufe his Soul into the Body of his Friend.

The Younger of thofe two Heroes expired upon the dead Body of his Friend; and you muft know, that from the Corruption of their Trunk, which without doubt, impregnated the Earth, two young Shrubs were feen to fprout out from amongft their dry bones; whofe Stem and Branches mingling promifcuoufly together, feemed to haften to grow only, that they might be twifted into a clofer Contexture.

It

It was visible that they had changed their Being, without forgetting what they had been; for their perfumed Buds leaned one upon another, and interchanged the Warmth of their Breathing, as it were to make themselves blow the sooner: But what shall I say of the loving Distribution, that maintained their Society? The Juyce, wherein the nourishment resides, never offered it self to their Stock, but they ceremoniously divided it: The one was never ill fed, but the other decayed for want; they both Suckt inwardly the Breasts of their Nurse; as ye Men do outwardly the Teats of yours. At length these happy Lovers brought forth Apples, but such miraculous Apples, as wrought greater Wonders than their Sires had done. All that eat of the Apples of the one, were instantly smitten with a Passionate Love for every one, that had tasted the Fruit of the other; and this happened almost daily, because all the Boughs of *Pylades* environed, or were environed by *Orestes*; and their Fruit, that were almost Twins, could not endure to be distant one from the other.

Nature however so cautiously distinguished their double Efficacy, that when one Man did eat the Fruit of one of these Trees, and another Man the Fruit of the other, it produced Reciprocal Friendship; and
when

when the same happened to two Persons of different Sexes, it begot Love; but such a vigorous Love, as still retained the Character of its Cause; for though that Fruit proportioned its effect to the Capacity, softening its force in a Woman, yet it still reserved somewhat that was masculine.

It is also to be remarked, That he of the two who had eaten most, was also most beloved. The Fruit was not only very lovely, but very sweet also; there being nothing so lovely and pleasant as Friendship. And indeed, it was the two qualities of Lovely and Good, which seldom meet in one subject, that put it into Vogue. How often by its miraculous virtue, hath it multiply'd the Examples of *Pylades* and *Orestes:* Since that time there have been instances of such as *Hercules* and *Theseus*, *Achilles* and *Patroclus*, *Nisus* and *Euryalus*; in short, of an infinite number of those, who by more than humane friendships, have consecrated their Memory in the Temple of Eternity. Cyens of these Trees were carried to *Peloponnesus*; and the place of Exercise, where the *Thebans* trained their Youth, was adorned with them: They were planted there in a streight line; and in the season, when the Fruit hung upon the branches, the youth who daily went into the place, being tempted by the beauty ty thereof, refrained not from eating; which according to custom presently influenced their Courage. They forthwith interchanged their Souls; every one becoming the half of another, living less in himself than in his Friend; and the faint-heartedest became bold and rash, for the sake of him he loved.

That celestial Passion warmed their Blood with so noble an Heat, that by the advice of the wiser, these Lovers were listed for the Wars into one Company: They have been called since, because of the Heroick Actions which they atchieved, *The sacred Band*. These Exploits went a great deal further, than the *Thebans* had promised themselves; for in time of Fight, every one of these Bravos, ventured such incredible Efforts, for the safety of his Lover, or for meriting his Affection, that the like hath not been seen in Antiquity: And indeed, so long as that amorous Company subsisted, the *Thebans*, who before were reckoned the worst Souldiers of all the *Grecians*, fought and still overcame afterwards the *Lacedemonians* themselves, the most War-like people upon the face of the Earth.

But amongst a vast number of laudable Actions caused by these Apples; they also produced (though innocently) some very ignominious ones.

Myrrah,

Myrrha, a young Lady of Quality, eat of them with *Cinyras* her Father; unfortunately the one was of *Pylades*, and the other of *Orestes*. Love immediately thereupon swallowed up, and so confounded Nature, that *Cinyras* could swear, I am my own Son-in-Law; and *Myrrha*, I am my own Step-Mother. In short, I think it sufficient to inform you of the nature of that Crime, that at Nine Months end the Father became the Grandfather of those he begot; and the Daughter was brought to bed of her Brothers.

Nor was Chance yet satisfied with this Crime; it so ordered matters, that a Bull coming into the Gardens of King *Minos*, unluckily found under a Tree of *Orestes* some Apples, which he swallowed down; I say unluckily, because the Queen *Pasiphae* daily eat of that Fruit: And so you have them mad in Love one with another. I shall not, however, speak of the enormous Enjoyment; it shall suffice to say, That *Pasiphae* plunged her self into a Crime, that was never matched before.

Exactly about that time, the famous Carver *Pigmalion*, was cutting a Marble Statue of *Venus* in the palace. The Queen, who loved good Workmen, made him a Present of a couple of these Apples: He eat the fairest; and because accidentally he wanted

wanted Water, which, as you know, is necessary for the cutting of Marble; he moistened his Statue with the Juyce of the other. The Marble, influenced by this Juyce, by little and little grew soft; and the efficacious virtue of that Apple, acting according to the design of the Workman, delineated within the Image, the draughts that it had met with in the Superfice; for it dilated, heated, and coloured, proportionably to the nature of the Places that it found in its passages. In fine, the Marble becoming animate, and being touched with the Passion of the Apple, embraced *Pigmalion*, with all her heart; and *Pigmalion*, transported with a reciprocal Love, took her for Wife.

In the same Province the young *Iphis* had eaten of that Fruit, with the fair *Ianthe*, her Companion in all the Exercises that are requisite to cause a reciprocal Friendship: Their Banquet was attended by its usual effect: But because *Iphis* had found it to be very agreeable to her Palate, she fed so heartily, that her Friendship encreasing with the number of Apples, wherewith she could not be satisfied, usurped all the functions of Love; and that Love growing still and still stronger, became more masculine and vigorous: For seeing her whole Body impregnated with that Fruit

Fruit, strove to form Motions that might answer the Caprices of her Will; it stirred up so powerfully its own matter, that it made to it self much stronger Organs, fit to comply with her Desire, and to satisfie her Love in its most manly Extent: My meaning is, That *Iphis* became what one ought to be, that marries a Wife.

I should term this strange accident a Miracle, had I still a name to give to the following Prodigy.

A most accomplished Youth, called *Narcissus*, had by his Love merited the Affection of a very lovely Maid, whom the Poets have celebrated by the Name of *Echo*. But seeing Women, as you know, are fonder of being much made of, than those of our Sex; she having heard the virtue of the Apples of *Orestes* much talked of, used means to procure a great many of them, from several parts; and because she apprehended, (Love being always fearful) That those of the one Tree, might have less force than the other, she would have him to taste of both: But so soon as he had eaten them, the image of *Echo* was quite blotted out of his memory; all his Love turned towards him who had digested the Fruit; and he became both the Lover and the Beloved; for the substance drawn from the Apple of *Pylades*, embraced within him

him the substance of the Apple of O-*restes*. That twin-fruit diffused through the whole mass of his blood, excited all the parts of his body to caress one another: His Heart, where their double virtue boiled, darted its flames inwards; all his Members animated with his Passion, endeavouring to penetrate one into another: Nay, not so much as his Image still burning in the cold Fountains, but attracted his Body to join it: In a word, poor *Narcissus* fell desperately in Love with himself.

I will not be tedious in relating to you his deplorable Catastrophy; the Ages of Antiquity have spoken enough of that: And besides I have Two Adventures still to acquaint you with, which will take up the time far better.

You shall know then, that the fair *Salmacis* frequented the company of the Shepherd *Hermaphroditus*, but with no other Privacies, than what the Neighbourhood of their Houses could allow of : When Fortune, who delights to disturb the most quiet and harmless Lives, so ordered, that in an Assembly of Plays, where the rewards for Beauty and Running were two of these Apples, *Hermaphroditus* gained that of the Race, and *Salmacis* the other of Beauty. Though they had been gathered together,

yet

yet it was from different Branches; because these amorous Fruits mingled together so cunningly, that one of *Pylades* was never without another of *Orestes*; and that was the reason, why appearing to be Twins, they plucked always a Couple at a time. The fair *Salmacis* eat her Apple, and pretty *Hermaphroditus* lockt his up in a Cupboard. *Salmacis* being inspired with the effects of her own Apple, and of that of the Shepherd, which began to grow hot in his Cupboard, felt her self attracted towards him, by the Sympathetick Flux and Reflux of the two.

The Shepherds Parents, who perceived the Amours of the Nymph, finding their advantage in that Alliance, endeavoured to entertain and promote it: And therefore having heard much talking of the Twin-Apples, as of a Fruit whose Juyce inclined People to Love, they distilled some of them; and having rectified the Spirit to the highest degree, found a means to make their Son and his Lover drink of it. The virtue of the Juyce, being sublimed to the highest degree it could be raised to, kindled in the Hearts of the Lovers, so vehement a desire of Conjunction, that at first sight *Hermaphroditus* was swallowed up in *Salmacis*, and *Salmacis* melted away in the Arms of *Hermaphroditus*: The one past into the other,

other, and of two of different Sexes, they made up I know not what double Person, that was neither Man nor Woman. When *Hermaphroditus* had a mind to enjoy *Salmacis*, he found himself to be the Nymph; and when *Salmacis* desired to be embraced by *Hermaphroditus*, she perceived her self to be the Shepherd. This couple though still retained its Unity; it Begat and Conceived, and yet was neither Man nor Woman: In short, in it Nature hath shewn a Miracle, which she hath never been able since to hinder from being One.

Well now, are not these pretty surprising Stories? Really they are; for to see a Daughter couple with her Father; a young Princess glut her self with the Amours of a Bull; a Man aspire to the Emjoyment of a Stone. Another to espouse himself; a Maid to Celebrate a Marriage, which she consummated as a youth; to cease to be a Man, without beginning to be a Woman; to become a Twin out of the Mothers Womb; and the Twin of another who had no Relation to him.

These are things quite out of the common Road of Nature; and nevertheless, you'll be more surprised at what I am about to tell you.

Amongst the sumptuous Variety of all sorts of Fruits and Trees, which were brought
from

World of the Sun.

from distant Climates, for the Marriage-Feast of *Cambyses*, there was presented to him a Cien of *Orestes*, which he caused to be grafted upon a Plane Tree; and amongst the Dainties of the last course, some Apples of the same Tree were served up to him.

The delicacy of the Dish invited him to eat heartily of it; and the substance of that Fruit, being after the three Concoctions converted into a perfect Seed; it formed in the Womb of the Queen, the Embryo of his Son *Artaxerxes*; for all the particulars of his Life, have made Physicians conjecture, that he must needs have been produced after this manner.

When the young Heart of that Prince, was old enough to deserve the anger of Love, it was not observed, that he sighed at all after any of his own kind: he loved nothing but Trees, Groves and Woods; but above all those that affected him, the lovely Plane Tree, whereon his Father *Cambyses* had formerly caused that shoot of *Orestes* to be graffed, wun his greatest affection.

His Constitution suited so nicely with the progress of the Plane Tree, that he seemed to grow with the Branches of it: He daily went and embraced it; in his Sleep he dreamt of nothing else; and under the Canopy of its Green Hangings, he dispatched all his Affairs. It was easily perceived,

that

that the Plane Tree smitten with a reciprocal Flame, was ravished with his Caresses: For on all occasions, without any apparent reason, its Leaves were seen to shake, and in a manner leap for Joy; the Branches bend round about his Head, as it were to make a Crown for him, and to reach down so near to his Face, that it was easie to be known, that it was rather to kiss him, than out of any natural inclination, of bending downwards. Nay, it was also observed, that out of Jealousie it ranked its Leaves in order, joining one close to the other, for fear least the Sun-Beams piercing through, might kiss him as well as it. The King on his part, set no more bounds to his Love; he had his Bed made under the Plane Tree, and the Tree not not knowing how to repay his Friendship, bestowed upon him the most precious thing that Trees have, which was its Honey-dew, that every Morning dropt upon his Face.

Their Caresses would have lasted longer, had not Death, the Enemy of Noble Actions, put an end to them: *Artaxerxes* died of Love in the embraces of his dear Plane Tree; and the Persians extreamly afflicted at the death of so good a Prince, resolved that they might give him satisfaction even after his Death, that his Body should be

burnt

burnt with the Branches of that Tree, and no other Wood employed in Confuming it.

When the Funeral Pile was kindled, the Flame was feen to twift it felf with that of the Fat of the Body; and their burning Locks which curled one into the other, to taper into a Pyramide as far as could be difcerned.

That pure and fubtile Fire divided not; but when it arrived at the Sun, whither you know all igneous matter tends, it formed the fprout of the Apple-Tree of *Oreftes*, which you fee there on your Right Hand.

Now the Breed of that Fruit is loft in your World, and I'll tell you how that mifchance happened.

Fathers and Mothers, who as you know, are only guided by intereft, in the management of their Domeftick Affairs, being vext that their Children, fo foon as they had eaten of thefe Apples, fquandered away upon their Friends all that they had, burnt all the young Plants they could find of that Tree; fo that the kind being loft, is the reafon why no true Friend is now to be found.

As faft then as thefe Trees were confumed by the Fire, the Rain that fell, calcined their Afhes, fo that the congealed Juyce

was

was petrified in the same manner, as the sap of burnt Fern is changed into Glass. Hence it is, that in all Climates of the Earth two Metallick Stones are formed of the ashes of those Twin-Trees, that now adays are called the Iron and Load-stone, which because of the Sympathy of the Fruits of *Pylades* and *Orestes*, the virtue whereof they have still retained, always aspire to embrace one another; and observe that if the piece of the Load-stone be the bigger, it attracts the Iron; or if the piece of Iron exceed in quantity, it attracts the Load-stone; as formerly it happened in the miraculous Effects of the Apples of *Pylades* and *Orestes*, of the one of which whosoever had eaten most, was the most beloved of him who had eaten the other.

Now Iron feeds so visibly upon the Load-stone, and the Load-stone upon the Iron, that the one rusts, and the other loses its force; unless they be put together for the reparation of what substance they lose.

Have you never observed a piece of Load-stone, laid upon the File-dust of Iron, you'll see the Load-stone cover it self in a trice with these metallick Atoms; and the amorous Heat wherewith they cling together, is so sudden and impatient, that when they have embraced one another in all places, you would say that there is not one grain of the Load-ſtone, that would not kiſs a grain of the Iron, nor a grain of the Iron, that would not be united to a grain of the Load-ſtone; for the Iron or Load-ſtone being ſeparated, continually ſend out from their Maſs, ſome moſt agile little Bodies, in queſt of that which they love: But when they have found that, having got their deſire, every one puts an end to their Progreſs; and the Load-ſtone takes its reſt in poſſeſſing the Iron, as the Iron wholly contents its ſelf in the enjoyment of the Load-ſtone. From the Sap then of theſe two Trees, the humour which hath given Being to thoſe two Metals has been derived.

Before that they were unknown; and if you have a mind to know, of what matter Arms were made for the War; *Sampſon* armed himſelf againſt the *Philiſtines*, with the Jaw-bone of an Aſs; *Jupiter* King of *Crete*, with Artificial Fire-works, whereby he imitated the Thunder, in ſubduing of his Enemies; and, in a word, *Hercules* with a Club overcame Tyrants, and cruſhed Monſters. But theſe two Metals, have another more ſpecifick relation to our two Trees: You muſt know, that though that Couple of Life-leſs Lovers incline towards the Pole, yet they never tend thither but in Company; and I'll tell you the Reaſon of it, it, after I have difcourfed to you a little about the Poles.

The Poles are the Mouths of Heaven, by which it fucks up again the Light, Heat, and Influences that it hath fhed upon the Earth: Otherwife if all the Treafures of the Sun, remounted not to their fource, (all its Brightnefs being only a duft of inflamed Atoms, which are detached from its Globe;) it would have been long ago extinguifhed, and fhone no more: Or that abundance of little igneous Bodies, heaping together upon the Earth, when they could not get out again, would have already confumed it. There muft then, as I have told you, be breathing Holes in Heaven, by which the Repletions of the Earth are difcharged, and others by which Heaven may repair its loffes; to the end the eternal Circulation of thefe little bodies of Life, may fucceffively pafs through all the Globes of this vaft Univerfe. Now the breathing holes of Heaven are the Poles, through which it retakes the Souls of all that die in the other Worlds without it; and all the Stars are its Mouths, and the Pores through which again it exhales its Spirits. But to fhew you, that this is not fo new an Imagination; when your Ancient Poets, to whom Phylofophy difcovered the moft hidden fecrets of Nature, fpake of an Hero,

whofe

whose Soul they would have said was gone to live with the Gods; they expressed it in this manner: He is gone up to the Pole, he is seated on the Pole, he hath past through the Pole; because they knew that the Poles where the only Avenues, through which Heaven receives again, all that is gone out from thence. If the Authority of these great Men be not sufficient to convince you, the Experience of your modern Navigators, who have sailed towards the North, may, perhaps, give you satisfaction. They have found, that the nearer they drew towards the Bear, during the Six Months of Night, when it was thought that Climate lay under a black Darkness, a great Light cleared the Horizon, which could not proceed but from the Pole; because the more one drew near to it, and by consequence removed from the Sun, that Light became greater. It is very probable then, that it proceeds from the Beams of day, and a great heap of Souls, which as you know, are only made of Luminous Atoms, that are returning to Heaven by their wonted Doors.

 This being so, it is no difficult matter to comprehend, wherefore the Iron rubbed with the Load-stone, or the Load-stone rubbed with the Iron, turns towards the Pole; for they being an Extract of the Body of *Py-*
lades

Iades and *Orestes*, and having still retained the Inclinations of the two Trees, as the two Trees have those of the Two Lovers, they ought to aspire to be rejoined to their Soul ; and therefore they skip towards the Pole, through which they perceive that it hath mounted ; but with this Reserve still, that the Iron never turns that way, unless it be touched by the Load-stone, nor the Load-stone, unless it be rubbed with the Iron ; by reason that the Iron will not quit a World, leaving his Friend the Load-stone behind, nor the Load-stone leaving its Friend the Iron, and that the one cannot resolve to perform this Voyage without the other.

This voice, as I think, was about to go on with another Discourse ; but the noise of a great Alarm that happened hindred it : All the Forest in an uproar, resounded with nothing but these Words, *The Plague, the Plague, stand upon your Guard, look about ye.*

I adjured the Tree, that had so long entertained me in discourse, to tell me the Cause of so great a Disorder. Friend, said he to me, we are not in these quarters, sufficiently as yet informed of all the Particulars of the Evil : I'll only tell you in Three Words, that the Plague wherewith we are threatned, is that which Men call a Fire;
we

we may very well call it so, because amongst us there is no such contagious Distemper. The remedy we are about to use against it, is to force our breath, and blow altogether, towards the place from whence the Inflamation comes, to the end we may drive back that bad Air. I believe that burning Feaver is occasioned us by a fiery Beast, that for some days has been roaming about our Woods; for seeing they never go without Fire, and cannot be without it, this, without doubt, is come to set some of our Trees on Fire.

We sent for the Animal *Frozen-nose*, to come to our Assistance; however is not as yet arrived. But, farewel, I have no time to talk, we must look to the publick Safety; nay, do you look to your self also, and fly for it, else you'll be in danger of being involved in our destruction.

I followed the counsel, but without much straining, because I knew my Legs. In the mean time I was so ill acquainted with the Geography of the Country, that at the end of Eighteen hours, I found my self at the back of the Forest that I thought I fled from; and to add to my fear, a hundred dreadful Thunder-claps stunned my Brains, whilst the ghastly and pale Glimpses of a Thousand flashes of Lightning put out my Eye-sight.

These

These Claps redoubled from time to time with so much fury, that one would have said, The Foundations of the World were about to be over-turned; and nevertheless the Heavens never appeared more serene. Though I was at my wits end, yet the desire of knowing the Cause of such an extraordinary Accident, made me go towards the place, from whence the noise seemed to proceed.

I had advanced about four hundred Furlongs, when I perceived in the middle of a great Plain, as it were, two Bowls, which having rustled and turned along time round one another, approached and then recoyled: And I observed that when they knocked one against the other, then were these great Claps heard; but going a little farther on, I found that what at a distance I had taken for two Bowls, were two Animals; one of which, tho round below, formed a Triangle about the middle, and his lofty Head with ruddy Locks, which floated upwards, spired into a Pyramide; his Body was bored like a Sieve, and through these little holes, that served him for Pores, thin flames glided, which seemed to cover him with a Plume of Fires.

Walking about there, I met with a very venerable old Man, who observed that famous conflict, with no less curiosity than my my self. He made me a sign to draw nigh, I obeyed, and we sat down by one another.

I had a design to have asked him the motive, that had brought him into that Country, but he stopt my Mouth with these words; Well then, you shall know the motive, that brought me into this Country. And thereupon he gave me a full account of all the particulars of his Voyage. I leave it to you to judge, in what amazement I was. In the mean while, to increase my consternation, as I was boyling with desire to ask him, what Spirit revealed my thoughts to him: No, no, cryed he, it's no Spirit that reveals your thoughts to me------

This new hit of Divination, made me observe him with greater attention than before, and I perceived that he acted my Carriage, my Gestures and Looks, that he postured all his Members, and shaped all the parts of his Countenance, according to the pattern of mine; in a word, my Shadow in relief could not have represented me better. I see, said he, you are in pain to know why I counterfeit you, and I am willing to tell you. Know then, that to the end I might know your inside, I disposed all the parts of my Body, into the same Order I saw yours in; for being in all parts scituated like you, by that disposition of matter, I excite in my self the same thought, that

M the

the same disposition of matter raises in you.

You will judge this to be a thing possible, if heretofore you have observed, that Twins who are like, have commonly the like Mind, Passions and Will: insomuch, that there were two Twins at *Paris*, who always had the same Sicknesses, and the same Health; married without knowing one anothers design, the same day and at the same hour; wrote Letters mutually to one another in the same Sense, Words and Stile; and in short, have upon the same Subject composed a Copy of the same kind of Verse, with the same Stops, Words and Order. Now don't you see, that it was impossible, but that the Composition of the Organs of their Bodies, being in all Circumstances alike, they must act in a like manner; seeing two like Instruments alike touched, ought to render a like Harmony? And that so I having conformed my Body wholly to yours, and become, if I may say so, your Twin; it is impossible, but that the same Agitation of Matter, must cause in both of us the same Agitation of Mind.

Having said so, he fell a counterfeiting me again, and thus went on:

You are at present in great pain to know, the Original of the Conflict of these two Monsters; but I will inform you of it. Know then, that the Trees of the Forest behind

hind us, being unable with their blowing, to repel the attempts of the fiery Beast, have had their recourse to the Animal *Frozen-Nose.*

I never heard of these Animals, said I to him, but from an Oak of this Country, and that in great haste too, because it was follicitous for its own safety; and therefore I would beg of you, to give me some account of them.

He thereupon spake to me in this manner: In this Globe where we are, we should see the Woods very thin sow'n, by reason of the great number of the fiery Beasts that destroy them; were it not for the Animals *Frozen-Noses,* which at the desire of the Forests their Friends, come daily to cure the Sick Trees: I say cure, for no sooner have they, from their Icy Mouth, blown upon the coals of that Plague, but they put it out.

In the World of the Earth, from whence both you and I are come, the fiery Beast is called the *Salamander*; and the Animal *Frozen-Nose,* is known by the name of *Remora.* Now you must know, that the *Remoras* live towards the extremity of the Pole, at the bottom of the *Mare Glaciale*; and it is the cold of these Fishes, evaporated through their Scales, which makes the Sea-Water in those quarters to freeze, though it be Salt.

Most Navigators, who have Sailed for the discovery of *Green-land*, have at length experienced, that in certain Seasons they found none of the Ice, which at other times had stopt them: Now though that Sea was open at the time, when it is bitterest Winter there, yet they have attributed the cause of it, to some secret Heat that had thawed it; but it is far more probable, that the *Remoras*, who only feed upon Ice, had at that time devoured the whole stock. Besides you are to know, that some Months after they have filled their Bellies, that strange Food of uneasy digestion, so chills their Stomack, that their very blowing of their Breath, freezes again all the Sea under the Pole. When they come on Land (for they live in both Elements) they fill their Paunch only with Hemlock, Wolf-bane, Opium and Mandrakes.

It's wondred at in our World, whence proceed those piercing North-Winds, that always bring Frost with them; but if our Country-men knew what we know, that the *Remoras* live in that Climate, they would know as well as we, that they proceed from a puff of their Breath, whereby they endeavour to blow back the heat of the Sun that draws near them.

That Stygian-Water wherewith the Great *Alexander* was poysoned, and whose
Cold-

Coldneſs petrified his Bowels, was the Piſs of one of theſe Animals. In fine, the *Remora* contains all the principles of Cold in ſo eminent a degree, that paſſing under a Ship, the Veſſel is ſeized with Cold, and ſtruck with ſuch a Numneſs, that it cannot wag out of the place. And that's the reaſon that one half of thoſe, who have cruiſed North-ward, for the diſcovery of the Pole, never came back again; becauſe it is a Mirracle if the *Remoras*, who are ſo numerous in that Sea, ſtop not their Veſſels. And ſo much for the Animals *Frozen-Noſes*.

But as to the Fiery Beaſts, they lodge on Land under Mountains of burning *Bitumen*, ſuch as *Ætna*, *Veſuvius* and others. The Pimples which you ſee upon the Breaſt of this Beaſt, that proceed from the Inflamation of his Liver, are------

Hear we put a ſtop to our Talk, that we might be more attentive to that famous Duel.

The *Salamander* attacked with much ardour; but the *Remora* defended impenetrably. Every daſh they gave one another, begot a clap of Thunder; as it happens in the Worlds there abouts, where the Claſhing of a hot Cloud with a cold, cauſes the ſame Report.

At every glance of Rage which the *Salamander* darted againſt its Enemy, out of its Eyes flaſhed a reddiſh Light, that ſeemed to kindle the Air in flying; it sweat boyling Oyl, and pissed *Aqua-fortis*.

The *Remora* on the other hand, that gross, square and heavy Animal, presented a Body scaled all over with Ysicles. Its large Eyes lookt like two Chrystal-plates, whose glances conveyed so chilling a light, that on what member of my Body it fixed them, I felt a shivering Winter-cold. If I thought to put my Hand before me, my Fingers ends were nummed; nay, the very Air about infected with its quality, condensed into Snow, the Earth hardned under his Steps; and I could reckon the Footings of the Beast, by the number of the Chil-blanes, that welcomed me when I trode upon them.

In the beginning of the Fight, the *Salamander* by the vigorous activity of its first heat, had put the *Remora* into a Sweat; but at length that Sweat cooling again, glazed all the Plain with so slippery an Ennamel, that the *Salamander* could not get up to the *Remora* without falling. The Philosopher and I knew very well, that the trouble of falling and rising so many times, had made it weary; for these Thunder-claps so dreadful before, that proceeded from the shock he gave its Enemy, were no more now but the dull Sound of those little After-claps, which denote the end of a

Storm;

Storm; and that dull Sound, deadned by degrees, degenerated into a Whizzing, like to that of a hot Iron plunged into cold Water.

When the *Remora* perceived, that the Fight was near an end, by the Weakness of the shock which was hardly felt by it, it raised it self upon an Angle of its Cube, and with all its weight fell upon the Breast of the *Salamander*, with so good success, that the Heart of the *Salamander*, wherein all the rest of its heat was contracted, bursting, made so fearful a Crack, that I know nothing in nature to compare it to.

Thus died the Fiery Beast, under the lazy resistance of the Animal *Frozen-Nose*.

Sometime after the *Remora* was gone, we approached the place of Battel; and the old Man having daubed his Hands over with the Earth, on which it had walked, as a Preservative against burning, laid hold on the Dead Body of the *Salamander*. Give me but the Body of this Animal, said he, and I've no need for Fire in my Kitchen; for provided it be hung upon the Pot-hook, it will Boyl and Roast all that's laid upon the Hearth. As for the Eyes, I'll carefully keep them; if they were cleansed from the Shades of Death, you'd take them for *two little Suns*. The Antients of our World knew well what use to make of them; they called them burning-Lamps, and never hung them up but in the Pompous Monuments of Illuſtrious Perſons.

The *Moderns* have found ſome of them, by digging into theſe famous Tombs; but their ignorant Curioſity made them put them out, thinking to find behind the broken Membranes, the Fire which they ſaw ſhine there.

The old Man went on ſtill, and I followed him, liſtning very attentively to the Wonders he told me. But ſince I have been ſpeaking of the Fight, I muſt not forget the Diſcourſe which we had, concerning the Animal *Frozen-Noſe*.

I don't think, ſaid he to me, that you have ever ſeen a *Remora*; for they are Fiſh that never riſe to the brim of the Water; nay, ſeldom or never do they leave the Northern Sea: But without doubt you have ſeen a ſort of Animals, which in ſome manner may be reckoned of their kind. I told you juſt now, that that Sea which reaches towards the Pole, is full of *Remoras*, that ſpawn in the mud as other Fiſhes do. You muſt know then, that that Seed, the Extract of all their maſs, ſo eminently contains all its Coldneſs, that if a Ship paſs over it, the Ship contracts one or more Worms, which become Birds; whoſe Blood is ſo deſtitute of heat, that though they have Wings, yet they are reckoned amongſt

Fiſhes

Fishes: And so the Pope, who knows their Original, forbids them not to be eaten in Lent; and these are the Fowls which in *France* they call *Maquereuses*.

I marched on still without any other design than to follow him, but so glad that I had found a Man, that I durst not take my Eyes off of him; so afraid was I to lose my Man. Mortal Youth, said he to me, (for I well perceive, that you have not as yet paid the tribute, which we owe to Nature, as I have done,) so soon as I saw you, I discovered in your Face, somewhat that shews you to be curious and inquisitive. If I be not mistaken in the Shape and Conformation of your Body, you must be a *Frenchman*, and a Native of *Paris*. That City is the place, wherewith I ended my Misfortunes, which I had carried about with me all over *Europe*.

My name is *Campanella*, and I am a *Calabrian* by Nation. Since my coming into the Sun, I have spent my time in visiting the Climates of this great Globe, that I may discover the Wonders of them: It is divided as the Earth is, into Kingdoms, Republicks, States and Principalities; so that Four-footed Beasts, Fowl, Plants and Stones, every one have their own; and though some of these allow no entrance amongst them, to Animals of a strange kind, especi-

especially to Men, whom the Birds above all others mortally hate, yet I can travel over all without any danger; becaufe the Soul of a Philofopher, is made up of more fubtile Parts, than the Inftruments which might be made ufe of to torment it. I was by good luck in the Province of the Trees, when the diforders of the *Salamander* began thofe great Thunder-claps, that you muft have heard as well as I, which guided me to their Field of Battel, whither you came foon after; but I was upon my return to the Province of Philofophers------- What, faid I to him, are there Philofophers alfo then in the Sun? Are there, replied the good Man, yes, fure; and they are the chief Inhabitants of the Sun, and the very fame, whom Fame in your World doth celebrate with fo full Mouth. You may fhortly converfe with them, provided you have the Courage to follow me; for before Three Days be over, I hope to be in their City. I don't think you can poffibly perceive the manner, how thefe great Spirits are tranfported hither. No certainly, cried I, for could fo many others been hitherto fo blind, as not to find the way? Or that after our Death, we fall into the Hands of an Examiner of Spirits, who according to our Capacity grants or refufes us our freedom in the Sun?

No-

Nothing of that, replied the old Man: It's by a Principle of Similitude, that Souls attain to this mass of Light; for this World is made up of nothing else, but the Spirits of every thing that dies in the Circumambient Orbs, such as *Mercury, Venus,* the Earth, *Mars, Jupiter* and *Saturn.*

Thus, so soon as a Plant, a Beast or a Man expire, their Souls without extinction mount to its Sphere, just as you see the flame of a Candle points up thither, in spight of the Tallow that holds it by the Feet. Now all these Souls being united to the source of Day, and purged from the gross matter that pestered them, exert far more noble Functions than those of Growing, Feeling and Reasoning; for they are employed in making the Blood, and vital Spirits of the Sun, that great and perfect Animal: And therefore also, you ought not to doubt, but that the Sun acts by the Spirit, more perfectly far than you do; since it is by the heat of a Million of these Souls rectified, whereof his own is an Elixir, that he knows the secret of Life, that he influences the matter of your Worlds, with the power of Generation, and that he makes Bodies sensible that they have a Being; and, in short, that he renders himself, and all things else, visible.

Now it remains, that I should clear to you, why the Souls of Philosopers, do not essentially join to the mass of the Sun, as those of other Men.

There are three orders of Spirits in all the Planets, that is to say, in the little Worlds which move about this.

The grosser serve only to repair the Plumpness of the Sun, the subtile insinuate into the place of his Beams; but those of Philosophers, having contracted no Impurity in their exile, arrive entire in the Sphere of Day to become its Inhabitants. Now they are not as others, a constituent part of its Mass; because the matter that composes them, in the point of their Generation, is so exactly mixed, that nothing can again dissolve it: Like to that which forms Gold, Diamonds, and the Stars, whereof all the parts are so closely interwoven and knit together, that the strongest Dissolvent cannot separate the Mixture.

Now these Souls of Philosophers, are so much in regard of other Souls, what Gold, Diamonds, and the Stars are, in respect of other Bodies; that *Epicurus* in the Sun, is the same *Epicurus*, who heretofore lived in the Earth.

The pleasure which I received in hearing that great Man, shortned my way; and I often started curious Questions, about which

which I importuned his opinion, that I might be thereby instructed: And really I never found so great goodness in any Man, as in him; for though by reason of the Agility of his Substance, he might in a few Days have arrived in the Kingdom of Philosophers; yet he chose rather to take the trouble of Jogging on with me, than to leave me amidst vast Solitudes.

Nevertheless he was in great haste; for I remember that having asked him, why he returned before he had surveyed all the Regions of that great world? He made answer, that his Impatience to see one of his Friends, who was newly arrived, obliged him to break off his Travels. I found by the sequel of his discourse, that his Friend was that famous Philosopher of our time *Monsieur des Cartes,* and that he made all haste to meet him.

He made answer also, when I asked him, what he thought of his natural Philosophy? that it ought to be read with the same respect, as Men listen to Oracles. Not, added he, but that the Science of natural things hath need, as other Sciences have, to prepossess our Judgment with *Axioms,* which it proves not: But the Principles of his are simple and so natural, that being once supposed, there is nothing that more necessarily satisfies all Appearances.

I could not forbear to interrupt him in this place: But methinks, said I to him, that that Philosopher hath always impugned the *Vacuum*: And nevertheless, though he was an *Epicurean*, yet that he might have the honour of giving a Beginning to the Principles of *Epicurus*, that's to say, to *Atomes*; he hath supposed for the beginning of things, a Chaos of matter throughly solid, which God divided into an innumerable number of little Squares, to every one of which he gave opposite Motions. Now he will have these Cubes, by rubbing one against another, to have crumbled themselves into pieces of all sorts of Figures: But how can he conceive, that these square Peices, could begin to turn separately, without granting a Vacuity betwixt their Angles? Must there not be necessarily a Void in the spaces, which the Angles of these Squares were forced to leave, that they might move? And then could these Squares, which only possessed a certain Extent before they turned, move in a Circle, unless in their Circumference they had possessed as much more? Geometry tells us, That that cannot be; one half then of that space, ought necessarily to have remained void, seeing there were as yet no *Atomes* to fill it.

My Philosopher made me answer, That *Monsieur des Cartes* himself would give us

a

a reason for that; and that being an obliging Gentleman, as well as a Philosopher, he would certainly be overjoyed to find a mortal Man in this World, that he might clear him of an Hundred Doubts, which his unexpected Death had constrained him to leave in the Earth, that now he had forsaken: That he did not think though, there was any great difficulty to answer that objection, according to his Principles which I had not examined, but as far as the weakness of my Wit could permit me; because, said he, the Works of that great Man, are so full and so subtile, that to understand them, there is need of the attention of the Soul of a true and consummated Philosopher: Which is the reason, that there is not a Philosopher in the Sun, but has a Veneration from him, insomuch that they will not dispute him the Precedency, if his modesty suffer him to take it.

To ease the trouble that the length of this Journey may give you, we will discourse of his Philosophy according to his Principles; which undoubtedly are so clear, and seem so abundantly satisfactory, through the admirable Wit of that great Genius, that one would say, He had assisted in the lovely and magnificent Structure of this Universe.

You

You remember, he saith, that our Understanding is Finite; so Matter being divisible *in infinitum*, it is not to be doubted, but that is one of the things, that it can neither imagine nor conceive; and that it is far above the reach of the Intellect, to give you a Reason for it: But, said he, though that cannot fall under the Senses, yet we conceive that it is so, by the knowledge we have of Matter; and we ought not, said he, suspend our Judgments about things that we conceive. Can we imagine the manner how the Soul acts upon the Body? Nevertheless, that is a truth not to be denied, nor doubted of; whereas it is a far greater absurdity to attribute to a *Vacuity*, that quality of yeilding to a Body, and that Space, which are the dependances of an Extent, which can only agree to a Substance; seeing by so doing one would confound the Notion of *Nothing* with that of a *Being*, and give Qualities to that, which can produce nothing, and cannot be the Author of any thing whatsoever. But, poor Mortal, said he, I perceive that these Speculations are tedious to thee; because as that Excellent Man saith, Thou hast never taken pains enough, to purifie thy Spirits from the mass of thy Body; and because thou hast rendred it so lazy, that it will perform no Functions now, without the aid of Senses.

I was about to reply, when he pulled me by the Arm, to shew me a Valley of wonderful Beauty. Do you perceive, said he to me, that bottom we are going down into? One would say, that the tops of the little Hills that bound it, were purposely Crown'd with Trees, that by the cool of their Shade, they would invite Travellers to repose.

At the foot of one of these Hills, the Lake of Sleep takes its source; it consists only of the Liquor of five Fountains, and if it mingled not with Three Rivers, and by its weight dulled the stream of their Waters, no Animal of our World could sleep. I cannot express how impatient I was to question him, about these Three Rivers, which I had never heard of before; but I was satisfied when he told me, that I should see all.

Soon after we arrived in the Valley, and much about the same time, upon the Carpet that borders that great Lake.

The truth is, said *Campanella* to me, you are happy, in that you see before you die, all the wonders of this World; it's a blessing for the Inhabitants of your Globe, that it hath produced a Man, who can inform them of the marvels of the Sun; seeing without you, they were in danger of living in gross Ignorance, and of tasting a thousand

Pleasures, without knowing whence they come; for it cannot be imagined, how liberally the Sun bestows his Largesses, upon all your little Globes: and this Valley alone, diffuses an infinite nnmber of Blessings, throughout the whole Universe, without which you could not live, nor so much as see the Day: Methinks that the sight of this Country alone, is enough to make you confess, that the Sun is your Father, and that he is the Author of all things. These Five little Rivers, that come and discharge themselves in it, run not above Fifteen or Sixteen hours; and nevertheless they seem to be so weary when they arrive, that hardly can they move; but they express their Lassitude by very different effects, for that of Sight contracts its self proportionably, as it approaches to the Lake of Sleep. The Hearing at its Mouth, confounds, wanders, and looses it self in the Pool: The Smelling raises a murmur, like that of a Man who snores: The Taste growing wallowish by the way, becomes altogether insipid: And the Feeling, so powerful a little before, that he lodged all his Comrades, is fain to conceal his own abode. On his part the Nymph of Peace, who resides in the middle of the Lake, with open Arms receives his guests, lays them in her Bed, and dandles them so gingerly, that to make them sleep,

she

she her self takes the pains to rock the Cradle. After they have for some time been thus confounded in this vast Bason, they again divide themselves at the further end, into Five Rivulets, which resume the same names when they issue out, that they left when they entered: But those which hasten most to be gone, and tug their Companions to set out, are the Hearing and Feeling; for the other Three wait, till these awaken them, and of all the rest the Taste lags always hindmost.

The Lake of Sleep is Vaulted over, with the black Arch of a *Grotto*. A great many Tortoises march slowly about the Shore; the Flowers of a Thousand Poppies, by looking into the Water, communicate to it its drouzy virtue. Not so much as *Dormice*, but come above Fifty Leagues to drink in it; and the purling of the Stream is so charming, that it would seem to break upon the Pebbles with Cadence, and to endeavour to compose Soporifick Musick.

The Wise *Campanella* without doubt foresaw, that I was about to feel the effects of it, and therefore he advised me to mend my pace. I would have obeyed him, but the Charms of that Water had so enveigled my Reason, that hardly could I understand his last Words. Sleep on, then, sleep on, I give you leave, said he; and indeed the Dreams that one has here are so perfect, that you'll be glad one day to call to mind, that which you are about to have. In the mean time, I'll divert my self in viewing the Rarities of the place, and then come back to you again. I think he talked no more, or at least the Vapours of sleep, had already put me out of condition, of being able to hear him.

I was in the middle of the learnedest and best conceived Dream that ever was, when my Philosopher came to awake me. I'll tell it you, when I can without digression; for it is very important you should know it, to let you see, with what freedom the Mind of the Inhabitants of the Sun acts, whilst Sleep captivates their Senses. For my part, I think, that that Lake evaporates an Air, which hath the property of depurating the Mind, entirely from the Fogs of the Senses; for nothing is presented to your thoughts, which does not seem to perfect and instruct you; and that's the reason why I highly respect those Philosophers, that are called Dreamers, who are laught at by the ignorant.

I opened then my Eyes with a start: I fancy I heard him saying; Mortal, you have slept enough, rise, if you would see

a

a Rarity that can never be imagined in your World. During the space of an hour, or thereabouts, since I left you; I have been walking by the Five Fountains, which come out of the Lake of Sleep. You may believe, that I have considered them with a great deal of Attention; they bear the name of the Five Senses, and glide very near one to another: That of the Sight seems to be a forked Pipe, full of the Powder of Diamonds, and little Looking-Glasses, that steal away and restore the Image of whatever presents; in its course it incompasses the Kingdom of *Linx* : That of the Hearing is in like manner double; it turns by its Insinuations like a *Dedalus*, and from the most hollow concavity of its Bed, one may hear an Eccho of all the noise that sounds round about: I am much mistaken, if they were not Foxes that I saw picking their Ears there: That of Smelling seems like the former, to divide it self into two Channels, hid under one and the same Arch; out of every thing it meets, it extracts somewhat invisible, whereof it composes a Thousand sort of Odours, which stand it in stead of Water; on the brink of that source, there are a great many Dogs, that rub and cleanse their Noses. That of the Taste runs by spurts, which commonly happen not above Three or Four times a

Day, and for that too a large van of Coral must be raised, and underneath that a great many little ones of Ivory; its Liquor resembles Spittle: But as to the Fifth, that of Feeling, it is so large and deep, that it environs all its Sisters, nay, and lays it self out at length in their Channels, and its thick Juyce, sheds it self abroad upon the green Turff, covered with sensitive Plants.

Now you must know, that stunned with Veneration, I admired the mysterious Turnings of all these Fountains: When after a great walk, I came to the entry where they discharged themselves into Three Rivers: But follow me, you'll better conceive the disposition of these things when you see them. A Promise that pleased me so well, throughly awoke me; I stretched out my Arm to him, and we kept the same way he had followed, walking along the Dykes that keep the Five Rivulets in their several Channels.

When we had gone about a Furlong, something as clear as a Lake presented it self to our Eyes. No sooner had the Wise *Campanella* perceived it, but he told me: At length, Son, we are got to the Port, I distinctly see the three Rivers.

I was so briskly transported with that news, that I thought I was become an Eagle.

Eagle. I flew rather than walked, and ran all about with so greedy a Curiosity, that in less than an hour, my Guide and I observed what now you shall hear.

Three great Rivers water the Fields of this Burning World: The First and largest is called Memory; the Second, narrower, but deeper, Imagination; and the Third, the last of the Three, is called Judgment.

Upon the Banks of Memory, one may hear continually a troublesome chattering of Jays, Parrots, Magpies, Starlings, Linnets, Chaffinches, and of all sorts of Birds, that chirp what they have learnt. In the Night time they are silent, for then they are taken up in feeding upon that thick Vapour, which exhales from these watery places; but their foul Stomack digests it so ill, that in the Morning, when they think it converted into their substance, it drops out of their Beak again, as clear as it was in the River.

The Water of that River seems to be clammy, and runs with much noise.

The Ecchos that are formed in its Caverns repeat the word, even to above a Thousand times: It breeds a kind of Monsters, who have a Face much like to that of a Woman. It hath others too more furious, who have a Square and Horned Head, not unlike to that of our Pedants. The whole Business of these is to cry, and nevertheless say no more but what they have heard one another say before.

The River of Imagination runs more gently; its light and shining Liquor sparkles on all hands: To look upon that Water like a Torrent of humide sparkles, one would think, that it observed no Order in its course. After I had considered it more attentively; I observed that the humour which flowed in its Channel, was of pure Potable Gold, and its froth of the Oyl of Talc. The Fish that it feeds are *Remoras*, *Syrenes* and *Salamanders*; instead of Gravel, it is full of those little Stones *Pliny* speaks of, with which Men become heavy, when they touch their wrong side, and light when they apply to them their Right side. I observed there also those other Stones, one of which *Giges* had in a ring, which render things Invisible; but above all, there are a great many Philosophers stones, which sparkle amongst its Sand. There were a great many Fruit-Trees upon the banks of it, especially those which *Mahomet* found in Paradise; their Branches swarmed with Phenixes, and I observed Crab-Stocks of that Tree, from which *Discord* pluckt the Apple which she threw amongst the three Goddesses; graffs of the Garden of the *Hesperides* had been graffed on them.

them. Each of these Two great Rivers, is divided into an infinite number of Branches, that are interlaced one with another; and I took notice, that when a great Rivulet of Memory, drew near to a less of Imagination, it immediately absorbed the other; but on the contrary, if the Rivulet of Imagination was the bigger, it dried up the Brook of Memory. Now seeing these Three Rivers, both in their Channels and Branches, run always by one another; wheresoever the Memory is strong, the Imagination diminishes; and this again swells, as the other is low.

Near to that the River of Judgment runs with an incredible slowness: It hath a deep Channel, its Liquor seems to be cold; and when it is shed upon any thing, it drys instead of moistening. In the Owze of its Channel grow Hellebore-Plants, whose Roots stretching out in long Filaments, even to the Mouth of it, purifie its Waters there: It breeds Serpents, and upon the soft grass that cover its banks, Thousands of Elephants repose themselves: It is divided, as the other two, into an infinite number of little Branches; it encreases as it advances in its course: and though it still gains ground, yet it continually ebbs and flows in it self.

All the Sun is watered by the Juyce of thefe Three Rivers; it ferves to fteep the burning Atomes of thofe that die in that great World; but this deferves very well to be handled more largely.

The Life of the Animals of the Sun is very long, and they expire not but by a natural Death, which only happens at the end of Seven or Eight thoufand Years; when by the continued Intenfion of mind, to which their fiery temper inclines them, the order of matter is jumbled; for in a Body, fo foon as Nature perceives, that it would require more time to repair the Antient Being, than to compofe a new one, fhe afpires to Diffolution; fo that the Animal may be feen daily not to Rot, but to fall into Particles like Red Afhes.

Death never happens but in this manner. The Animal then being expired, or, to fay better, extinct; the little igneous Bodies that made up his fubftance, enter into the grofs matter of this burning World, until Chance hath watered them with the Liquor of the Three Rivers; for then becoming moveable by their Fluidity, that they may quickly exert the Faculties, of which that Water hath given them an obfcure Knowledge, they faften together into threads, and by a Flux of Luminous points, fharpen themfelves into Beams, and then
disperfe

disperse into the Neighbouring Spheres; where they are no sooner wafted, but they themselves dispose the matter, as much as they can, into a Form proper for the exerting all the functions, whereof they have contracted an Instinct, in the Water of the Three Rivers, the Five Fountains, and the Lake; and therefore they suffer themselves to be attracted to Plants for Vegetation; the Plants suffer themselves to be brouzed upon by Animals for Sensation; and the Animals suffer themselves to be eaten by Men; that so being converted into their substance, they may repair the Three Faculties of Memory, Imagination and Judgment, of whose power the Rivers of the Sun, had given them a Fore-taste.

Now according as the Atomes have been more or less soaked, in the Liquor of these Three Rivers; they furnish Animals with more or less Memory, Imagination or Judgment; and according as in the Three Rivers, they have inbibed more or less, of the Liquor of the Five Fountains, and of the Lake, they form to them Senses more or less perfect, and produce Souls more or less drowzy.

This is in a manner what we observed, concerning the nature of these Three Rivers. Little scattered veines of them may be met with every where; but as for the
principal

principal Branches, they run with a ſtreight courſe to the Province of Philoſophers: And therefore we returned to the high way again, not leaving the Current wide of us, farther than it was neceſſary to get upon the Cauſey. We ſaw the Three great Rivers always running by our ſide; but for the Five Fountains, we beheld them turning and winding below in the Meadows. That's a very pleaſant Road, though it be ſolitary; the Air there is pure and thin, which nouriſhes the Soul, and makes it reign over the Paſſions.

At the end of Five or Six days Journey, as we were diverting our ſight, with the various and rich Proſpects of the Country, we heard a languiſhing Voice, like the groaning of a ſick Perſon. We drew near the place, from whence we judged it might come, and found upon the brink of the River Imagination, an old Man fallen backwards, who complained grievouſly. Tears of compaſſion came into my Eyes, and Pity obliged me to ask the poor wretch what he ailed. That Man, anſwered *Campanella*, turning towards me, is a Philoſopher reduced to Extremity: For we die oftner than once; and ſeeing we are but parts of this Univerſe, we change our form, that we may go Live elſewhere; which is not a Misfortune, ſince it is a

way

way to perfect ones Being, and to attain to an infinite number of Sciences: His distemper is that, which makes all great Men for the most part to die.

His Discourse obliged me to consider the Patient more attentively, and at the first glance, I perceived that his Head was as big as a Tun, and open in many places. Come, come, said *Campanella* to me, pulling me by the Arm, all the assistance that we may think we could give to this dying Man, would be unprofitable, and only trouble him the more. Let's Jog on, for indeed his Evil is Incurable: The Swelling of his Head, proceeds from the Restlessness of his Mind; for though the Ideas, wherewith he has filled the Three Organs, or the Three Ventricles of his Brain, be but very small Images; yet they are Corporeal, and by consequence capable of filling a great place, when they are very numerous. Now you must know, That that Philosopher hath so dilated his brain, by stuffing it with notion upon notion, that being unable longer to contain them, it hath burst. That way f dying is common to great Genies, nd it is called, To crack with Wit.

We marched on still discoursing; and hat presented first to our view, furnishd us with matter of Conversation. I ould have been very willing though, to
have

have left the obscure Regions of the Sun, and gone again into the Luminous; for the Reader must know, That all the Countries are not Diaphanous, there are some of them that are obscure, like those of our World; and which, were it not for the light of the Sun, that is perceived beyond them, would be covered with Darkness. Now proportionably as one enters into the obscure Regions, he insensibly becomes so himself; and in the same manner, when one approaches the transparent, he perceives himself stript of that somber Obscurity, by the vigorous Irradiation of the Climate.

I remember, that upon occasion of this earnest desire I had, I asked *Campanella*, if the Province of Philosophers was resplendent or darkish? It is more darkish than resplendent, answered he: For as we still Sympathize much with the Earth, our native Country, which of its own nature is Opacous; so we could not fit our selves in the clearer Regions of this Globe. Nevertheless by a vigorous bending of the Will, we can render our selves Diaphanous, when we have a mind to it: Nay, and most part of the Philosophers, do not speak with the Tongue, but when they have a mind to communicate their Thoughts, they purge themselves by the Ejaculations
of

of their Fancy, of a somber Vapour, under which commonly they keep their Conceptions covered; and so soon as they have remanded to its place, that obscurity of the Spleen which darkened them, seeing their Body is then Diaphanous, one may perceive through their Brain what they remember, what they imagine, what they judge; and in their Liver and Heart, what they desire, and what they resolve: For though these little Pictures be more imperceptible, than any thing that we can devise; yet in this World our Eyes are clear-sighted enough, easily to distinguish even the smallest Ideas.

Thus when any of us would discover to his Friend, the Affection he has for him, his Heart is perceived to dart out Beams, as far as his Memory, upon the Image of him he Loves: And when on the contrary, he would testifie his Aversion, his Heart is seen to Thunder against the Image of him he hates, storms of burning Sparks, and to retreat backward as far as it can: In the same manner when he speaks within himself, the Ideas are clearly to be observed, that's to say, The Characters of every thing he meditates upon, which by rising and falling, imprinting and effacing, present to the Eyes of the Beholder, not an articulated Discourse, but a History of all his thoughts in *taille-doux*. My

My Guide would have gone on, but he was diverted by an Accident, the like was never heard before: And that was, that all of a sudden we perceived the Earth blackened under our Feet, and the Heavens kindled before with Beams, extinguished over head, as if a Canopy Four Leagues broad, had been spread betwixt us and the Sun.

It would be no easie matter for me to express, what we imagined in that Juncture: All sorts of Terrors, even that of the Worlds end, seized us, and none of these Apprehensions seemed to us to be improbable; for to see night in the Sun, or the Air overcast with Clouds, is a Miracle that never happens there. And yet this was not all; for immediately after a sharp and skreaking noise, like to that of the winding up of a Jack, came to grate our Ears; and at the very same time a Cage fell at our Feet. No sooner had it rested upon the Sand, but it opened, and was brought to bed of a Man and a Woman; they had an Anchor with them, which they fastened to the Roots of a Rock; the next thing they did, was to make towards us. The Woman led the Man, and with threats dragged him forward. When she was come very near us, Gentlemen, said she, in some little disorder, is not this the Province of Philosophers? I made answer, No; but that

that we hoped to be there within the space of Four and twenty hours; and that the old Man, who allowed me his Company, was one of the chief Ministers of that Monarchy. Seeing you are a Philosopher, replied the Woman, addressing her self to *Campanella*, without going further, I must discharge my heart to you.

To tell you then in a few Words, the occasion of my coming hither, you must know, that I come to complain of a Murder committed on the person of the Youngest of my Children; the Barbarian, whom I hold here, hath twice kill'd him, though he be the Father. We were extreamly puzled at this Discourse, and therefore I desired to know, what she meant by a Child killed twice? Know, answered the Woman, that in our Country, amongst the other Statutes of Love, there is a Law regulates the number of Kisses, which a Husband is obliged to give his Wife: And it's for that reason, that every evening a *Physician*, within his own precinct, visits all the Houses, where having viewed the Husband and Wife, he taxes them for that night, according to their Health, strong or weak, in more or less Embraces. Now my Husband there was adjudged to Seven: Nevertheless, being netled at some angry words gave him, as we were going to Bed, he did not so much as touch me, all the while we were in bed: But God, who avenges the cause of the afflicted, permitted, That that Wretch being tickled in a dream, by remembring the Kisses, which he unjustly detained from me, let a Man be lost. I told you, that his Father hath killed him twice, because by hindering him to be, he is the cause that he is not, there is his first Murder; and he is likewise the cause, why he hath not been, there's his second: Whereas an ordinary Murderer knows very well, that he whom he destroys, is no more in being; but he cannot hinder, but that he hath had a Being. Our Magistrates would have rendred Justice in the matter; but the Crafty Man alledged for excuse, That he would have performed his conjugal Duty, had he not been apprehensive, that kissing me in the rage that I had put him into, he might have begot a Madman.

The Senate puzled at that Plea, ordered us to go and appear before the Philosophers, and plead our Cause there. So soon as we received the Order to be gone, we put or selves into a Cage, hung by the Neck of that great Fowl, which you see there; from whence by means of a Pully which we fastned to it, we let our selves down to the ground, and hoist our selves up up into the Air. There are people in our Province, purposely appointed to tame them when they are young, and breed them up to the work we employ them in. That which chiefly makes them tractable, contrary to their fierce nature, is, that to satisfie their unsatiable Hunger, we give them the Bodies of all the Beasts that die to feed on. After all, when we have a mind to sleep, (for because of the constant excesses of Love, which weaken us, we stand in need of Rest:) We let loose into the open Fields, at convenient distances, Twenty or Thirty of these Fowls, each tied to a rope, who taking flight with their great Wings, display in the Sky a Night larger than the Horizon. I was very attentive both to her Discourse, and in great extasie, to consider the prodigious bulk of that Giant-Bird: But so soon as *Campanella* had lookt a little upon it, Ha! verily, cried he, it is one of those Feathered Monsters called *Condores*, which are to be seen in the Isle of *Mandragora* in our World, and all over the Torrid Zone, they cover an Acre of ground with their Wings: But seeing these Animals grow Huger, according as the Sun, under which they are bred, is hotter in the World of the Sun, they must needs be of a prodigious Greatness.

However, added he turning to the Woman, you muſt of neceſſity accompliſh your Journey; for it belongs to *Socrates*, who hath the inſpection of Manners, to decide your Cauſe. In the mean time, I adjure you to tell us, what Country you are of, becauſe ſeeing it is but three or four years, ſince I arrived in this World, I am but very little as yet acquainted with the Map of it?

We are, anſwered ſhe, of the Kingdom of Lovers: That great State is on one ſide bordered by the Republick of Peace, and on the other, by that of the Juſt.

In the Country I come from, at Sixteen years of Age, Boys are put into the Novitiat of Love: It is a very ſtately Palace, that takes up almoſt a quarter of the City. The Maids are put into it at Thirteen; and both accompliſh their year of Probation there; during which the Boys are only employed, in meriting the affection of the Girls, and the Girls in rendring themſelves worthy of the Love of the Boys. When the Twelve Months are up, the faculty of medicine in Body, go and viſit this Seminary of Lovers: They feel them all over, one after another, even to the moſt Privy parts of their Body; make them couple before them; and then according as the Male, upon Tryal, is found to be vigorous and
well

well-shaped, they give him for Wives, Ten, Twenty, Thirty or Forty Maids, such as loved him; provided he reciprocally love them. The Husband, nevertheless, cannot lie but with Two at a time, and it is not lawful for him to Embrace any of them, so long as she is with Child. Such as are found to be Barren, are only employed in Service; and Men who are impotent are made Slaves, and may carnally mingle with the Female-Drudges. After all, when a Family hath more Children than it can bring up, the Republick takes care of them: But that's a misfortune that very seldom happens; because so soon as a Woman is brought to Bed in the City, the publick Treasury furnishes a yearly Pension for the Education of the Child, according to its Quality; which on certain days, the Treasurers of State themselves, carry to the House of the Father: But if you have a mind to know more, step into our Panier, it is big enough for Four. Seeing we are going the same way, we'll talk and make our Journey the shorter.

Campanella was of the mind, that we should embrace the offer; and I was likewise very glad of it, to avoid being tired: but when I came to help them to weigh their Anchor, I was much surprized to find, that instead of a great Cable, which ought to bear it up, it hung only by a Silken thread as small as a Hair. I asked *Campanella* how it could be, that a Mass so heavy as that Anchor was, did not by its weight break so weak a thing? And the good Man made answer, That that Line did not break, because being spun all of an equal bigness, there was no reason why it should sooner break, at one place than another. We all stowed our selves into the Pannier, and then hoisted up our selves by the Pully, as high as the Fowl's Throat, where we appeared no bigger than a Bead hanging at its Neck. When we were up as high as the Pully, we fastened the Cable by which our Cage hung, to one of its smallest Down-feathers, which nevertheless was as big as ones Thumb; and so soon as the Woman, had made a sign to the Bird to be gone, we perceived it cleave the Air with a violent Rapidity. The *Condore* hastned or slackened its flight, soared or stooped, according to its Mistresses pleasure, whose Voice served it for a Bridle. We had not flow'n Two hundred Leagues, when we perceived on the Earth, to the left Hand, a night like to that, which our living Umbrello made under us. We asked the stranger Woman, what she thought it might be? It's another Malefactor, answered she, who is going also to receive Justice in the Province, whither

ther we are going: His Fowl, without doubt, is ftronger than ours, or otherwife we have trifled away a great deal of time by the way, for he fet not out till after I was gone. I asked her, what Crime that poor Wretch was accufed of? He is not barely accufed, anfwered fhe; he is condemned to dye, becaufe he is already convicted of not being afraid of Death. How then, faid *Campanella* to her, do the Laws of your Country enjoyn Men to be afraid of Death? Yes, replied the Woman, they enjoyn all, except thofe who are admitted into the Colledge of the Wife; for our Magiftrates have found by fad Experience, that he who fears not to lofe Life, may take it from any Body elfe.

After fome other difcourfes that followed thefe, *Campanella* had a mind to make a larger enquiry into the Manners of her Country: He asked her then, what were the Laws and Cuftoms of the Kingdom of Lovers? But fhe begged his pardon, if fhe did not anfwer him; becaufe fince fhe was not born there, and knew them but in part, fhe was afraid, fhe might fay too much or too little. I came into that Province, continued the Woman; but I and all my Predeceffors, are originally of the Kingdom of Truth; my Mother was delivered of me here, and never had another Child; fhe brought me up in the Country, till I was Thirteen Years of Age, when the King by the advice of Physicians, commanded her to carry me to the Kingdom of Lovers, from whence I come; to the end, that having my Breeding in the Palace of Love, that Education which is more chearful and soft, than the Breeding of our Country, might render me more Fruitful than she had been. My Mother carried me thither, and placed me out into that House of Pleasure.

I had much ado to comply with their Customs: At first they appeared to me to be very rude; for, as you know, the opinions that we have suckt in with our Mothers Milk, seem always to us to be the most rational; and then I was but just come from the Kingdom of Truth, my native Country.

Not but that I perceived very well, that the Nation of Lovers, lived with more Condescension and Indulgence, than ours did; for though every one gave it out, That my Sight wounded dangerously, that my Looks killed, and that my Eyes glanced out Flames, which consumed Hearts; yet the Goodness of all, and especially of the Young Men, was so great, that they carressed, kissed and hugg'd me, instead of revenging the Evil that I had done them.

Nay

Nay, I was even vexed with my self, for the disorders that I was the cause of; and that was the reason, that out of Pity I told them one day, That I was resolved to run away. But alas! how can you save your self, cryed they all, embracing my Neck, and kissing my Hands: Your House is on all Hands beset with Water; and so great the danger appears to be, that undoubtedly you and we both had been already drowned, without a Miracle.

How, said I to our Historian, is the Country of Lovers then subject to Inundations? It may very well be said to be, replied she; for one of my Gallants (and that Man would not have deceived me, because he loved me) wrote to me, That for grief of my departure, he had shed an Ocean of Tears. I saw another who assured me, That within the space of three days, his Eyes had distilled a Fountain of Water: And as I was cursing, for their sakes, the fatal Hour when first they saw me, one who reckoned himself of the number of my Slaves, sent me word, that the night before, an overflowing of his Eyes, had caused a Deluge. I was about to have left the World, that I might no longer be the cause of so many Evils, had not the Messenger subjoined, that his Master had charged him to assure me, That I had no cause

cause to fear any thing, seeing the Furnace of his Breast, had dried up that Deluge. In fine, you may Conjecture how waterish the Kingdom of Lovers must needs be, since with them it is to weep but by halves; when from under their Eye-lids, there springs no more but Rivulets, Fountains and Torrents.

I was in great pain, what Machine I could find, to save my self out of all these Waters, that were like to over-whelm me: But one of my Lovers, who was called *The Jealous*, advised me to pluck out my Heart, and then embark in it; that I needed not fear, but that it would hold me, because it held so many others; nor that I should sink, because it was too light: That all I was to be afraid of, was to be burnt, because the Materials of such a Vessel, was much subject to Fire: That I should be gone then upon the Sea of his Tears; that the Fillet of his Love, would serve me for a Sail; and that the favourable Gale of his Sighs, in spight of his Rivals Storm, would carry me to Shoar.

I was a long while a musing with my self, how I could put that enterprise into execution. The natural Fearfulness of my sex, hindred me from daring; but at length the opinion that I had, that if the thing were not feasible, a Man would not
be

be such a Fool as to advise it, and far less a Lover to his Mistress, gave me the Boldness.

I snatched a knife, slit up my Breast; nay, with both my hands I was already searching in the wound, and with an undaunted look, I felt for my Heart to pluck it out, when a Young Man, who loved me, came in. In spight of me he wrested the Weapon from me, and then asked me the motive of that desperate Action, as he called it. I gave him an account of it; but was much surprized, when within a quarter of an hour after, I understood that he had brought the *Jealous* before the Justice. Nevertheless the Magistrates, who, perhaps, feared they might be biassed by the example or novelty of the Accident, referred that Cause to the Parliament of the Just. There he was Condemned, besides perpetual Banishment, to go end his Days as a Slave, in the Land of the *Republick* of *Truth*; with prohibition to all that should descend of him, to the Fourth Generation, ever to return into the Province of Lovers; nay, moreover he was enjoyned upon pain of Death, never more to use an Hyperbole.

Since that time I entertained a great affection for the Young Man that saved me; and whether it were for that good
Office,

Office, or because of the Passion, wherewith he served me, when my Novitiat and his were out, I did not refuse him, when he demanded me for one of his Wives.

We have always since lived very well together, and should have continued to do so still, had he not, as I have told you, killed one of my Children twice; for which I am going to emplore Justice, in the Kingdom of Philosophers.

Campanella and I were much astonished at the silence of that Man; and therefore I endeavoured to comfort him, judging, that such a profound Taciturnity, was the Daughter of a very deep Remorse: But his Wife took me off of that. It is not, said she, the excess of Sorrow that stops his Mouth, but our Laws forbid all Criminals, that stand Indited, to speak unless it be before their Judges.

During that conversation, the Fowl was going on still, but I was strangely amazed when I heard *Campanella*, with a Countenance full of transports of Joy, cry out: Now welcome the dearest of all our Friends: Let's go, Gentlemen, continued the good Man, Let's go meet *Monsieur Des Cartes*; come let us alight, he is just now arrived, and but Three Leagues of. For my part, I was exceedingly surprized at this Eruption, for I could not comprehend, how he could come to know the arrival of a Man, of whom we had received no News. Certainly, said I to him, you have juft now feen him in a Dream. If you call a Dream, said he, what your Soul can fee with as great a certainty, as your Eyes fee the light of Day; I confefs it. But, cried I, is it not a Ravery to think, that *Monfieur Des Cartes*, whom you have not feen, fince you left the World of the Earth, is now but Three Leagues off, becaufe you have imagined it to be fo?

I had juft uttered the laft Syllable, when we faw *Des Cartes* come. Immediately *Campanella* ran to embrace him: They talked together a long while; but I could not mind all the obliging Complements they made to one another, I was fo full of defire to learn of *Campanella* his Secret of Divination. That Philofopher, who read my Paffion in my looks, gave his Friend an account of it, and prayed him not to take ill if he fatisfied me. *Monfieur Des Cartes* anfwered with a fmile, and my learned Preceptor difcourfed in this manner. Out of all Bodies *Species*'s exhale, that's to fay, Corporeal Images, which dance in the Air. Now thefe Images ftill retain, notwithftanding their Agitation, the Figure, Colour, and all the other Proportions

ons of the Object, from which they proceed: But seeing they are very pure and subtile, they pass through our Organs, without causing the least Sensation in them: They penetrate into the Soul, where because of the Delicateness of its Substance, they imprint themselves, and so represent to it Objects very remote, which the Senses cannot perceive. It's a thing that commonly happens here, where the mind is not shut up in a Body, made of gross Matter, as in thy World. We'll tell thee how that comes to pass, when we have had the leisure, fully to satisfie the mutual Desire, that each of us have, to converse with the other; for certainly thou well deservest to be used with the greatest Civility.

FINIS.

ERRATA.

PAge 5. line *ult.* read *bought up.* p. 26. l. 1. r. *many.* p. 31. l. 26. r. *height.* p. 50. l. 4. r. *in.* p. 53. l. 14. dele *of it.* p. 100. l. 12. r. *directs.* p. 101. l. 29. r. *Croud.* p. 111. l. 25. r. *mildest.* p. 121. l. 29. r. *but.* p. 127. l. 21. r. *food.* p. 128. l. 2. r. *stunn.* p. 136. l. 18. add *is.* p. 169. l. 18. r. *wherein.* p. 175. l. 19. r. *for.* p. 183. l. 9. r. *least.*

are not so many Stones as clods of Earth, nor so many Animals as Plants, nor so many Men as Beasts; just so there ought not to be so many Spirits as Men, by reason of the difficulties that occur in the Generation of a perfect Creature.

I asked him, if they were Bodies as we are? He made answer, That they were Bodies, but not like us, nor any thing else which we judged such; because we call nothing a Body commonly, but what we can touch: That, in short, there was nothing in Nature, but what was material; and that though they themselves were so, yet they were forced, when they had a mind to appear to us, to take Bodies proportionated to what our Senses are able to know; and that, without doubt, that was the reason, why many have taken the Stories that are told of them, for the Delusions of a weak Fancy, because they only appeared in the night time: He told me withal, That seeing they were necessitated to piece together the Bodies, they were to make use of in great haste, many times they had not leisure enough, to render them the Objects of more Senses than one at a time, sometimes of the Hearing, as the Voices of *Oracles*, sometimes of the Sight, as the *Fires* and *Visions*, sometimes of the Feeling, as the *Incubusses*;
and

and that these Bodies being but Air condensed, in such or such a manner, the Light dispersed them by its heat, in the same manner, as it scatters a Mist.

So many fine things as he told me, gave me the curiosity to question him about his Birth and Death; if in the Country of the Sun, the *individual* was procreated by the ways of Generation, and if it died by the dissolution of its Constitution, or the discomposure of its Organs? Your senses, replied he, bear but too little proportion to the Explication of these Mysteries: Ye Gentlemen imagine, that whatsoever you cannot comprehend is spiritual, or that it is not at all; but that Consequence is absurd, and it is an argument, that there are a Million of things, perhaps, in the Universe, that would require a Million of different Organs in you, to understand them. For instance, I by my Senses know the cause of the Sympathy, that is betwixt the Loadstone and the Pole, of the ebbing and flowing of the Sea, and what becomes of the Animal after Death; you cannot reach these high Conceptions but by Faith, because they are Secrets above the power of your Intellects; no more than a Blind-man can judge of the beauties of a Land-skip, the Colours of a Picture, or the streaks of a Rain-bow; or at best he will fancy them to befomewhat palpable, to be like Eating, a Sound, or a pleafant Smell: Even fo, fhould I attempt to explain to you, what I perceive by the Senfes which you want, you would reprefent it to your felf, as fomewhat that may be Heard, Seen, Felt, Smelt or Tafted, and yet it is no fuch thing.

He was gone on fo far in his Difcourfe, when my Juggler perceived, that the Company began to be weary of my Gibberifh, that they underftood not, and which they took to be an inarticulated Grunting: He therefore fell to pulling my Rope afrefh, to make me leap and skip, till the Spectators having had their Belly-fulls of Laughing, affirmed that I had almoft as much Wit, as the Beafts of their Country, and fo broke up.

Thus, all the comfort I had during the mifery of my hard Ufage, were the vifits of this officious Spirit; for you may judge what converfation I could have, with thefe that came to fee me, fince befides that they only took me for an Animal, in the higheft clafs of the *Category* of Bruits, I neither underftood their Language, nor they mine. For you muft know, that there are but two Idioms in ufe in that Country, one for the Grandees, and another for the People in general.

That

World of the Moon.

That of the great ones is no more, but various inarticulate Tones, much like to our Musick, when the Words are not added to the Air: and in reality it is an Invention, both very useful and pleasant; for when they are weary of talking, or disdain to prostitute their Throats to that Office, they take either a Lute, or some other Instrument, whereby they communicate their Thoughts, as well as by their Tongue: So that sometimes Fifteen or Twenty in a Company, will handle a point of Divinity, or discuss the difficulties of a Law-suit, in the most harmonious Consort, that ever tickled the Ear.

The second, which is used by the Vulgar, is performed by a shivering of the Members, but not, perhaps, as you may imagine, for some parts of the Body signifie an entire Discourse; for example, the agitation of a Finger, a Hand, an Ear, a Lip, an Arm, an Eye, a Cheek, every one severally will make up an Oration, or a Period with all the parts of it: Others serve only instead of Words, as the knitting of the Brows, the several quiverings of the Muscles, the turning of the Hands, the stamping of the Feet, the contorsion of the Arm; so that when they speak, as their Custom is, stark naked, their Members being used to gesticulate their Conceptions,

tions, move so quick, that one would not think it to be a Man that spoke, but a Body that trembled.

Every day almost the Spirit came to see me, and his rare Conversation made me patiently bear with the rigour of my Captivity. At length, one morning I saw a Man enter my Cabbin, whom I knew not, who having a long while licked me gently, took me up in his Teeth by the Shoulder, and with one of his Paws, wherewith he held me up, for fear I might hurt my self, threw me upon his Back; where I found my self so softly seated, and so much at my ease, that being afflicted to be used like a Beast, I had not the least desire of making my escape; and besides, these Men that go upon all four, are much swifter than we, seeing the heaviest of them, make nothing of running down a Stagg.

In the mean time I was extreamly troubled, that I had no news of my courteous Spirit; and the first night we came to our Inn, as I was walking in the Court, expecting till Supper should be ready, a pretty handsome young Man came smiling in my Face, and cast his Two Fore-Legs about my Neck. After I had a little considered him: How! said he in *French*, do not you know your Friend then? I leave you to judge in what

case

case I was at that time; really, my surprise was so great, that I began to imagine, that all the Globe of the Moon, all that had befallen me, and all that I had seen, had only been Enchantment: And that Beast-man, who was the same that had carried me all day, continued to speak to me in this manner; You promised me, that the good Offices I did you, should never be forgotten, and yet it seems you have never seen me before; but perceiving me still in amaze: In fine, said he, I am that same *Demon of Socrates*, who diverted you during your Imprisonment, and who, that I may still oblige you, took to my self a Body, on which I carried you to day: But, said I interrupting him, how can that be, seeing that all Day you were of a very long Stature, and now you are very short; that all day long you had a weak and broken Voice, and now you have a clear and vigorous one; that, in short, all day long you were a Grey-headed old Man, and are now a brisk young Blade: Is it then that whereas in my Country, the Progress is from Life to Death; Animals here go Retrograde from Death to Life, and by growing old become young again.

So soon as I had spoken to the Prince, said he, and received orders to bring you

to Court, I went and found you out where you were, and have brought you hither; but the Body I acted in, was so tired out with the Journey, that all its Organs refused me their ordinary Functions, so that I enquired the way to the Hospital; where being come in I found the Body of a young Man, just then expired by a very odd Accident, but yet very common in this Country ——— I drew near him, pretending to find motion in him still, and protesting to those who were present, that he was not dead, and that what they thought to be the cause of his Death, was no more but a bare Lethargy; so that without being perceived, I put my Mouth to his, by which I entred as with a breath: Then down dropt my old Carcass, and as if I had been that young Man, I rose and came to look for you, leaving the Spectators crying a Miracle. With this they came to call us to Supper, and I followed my Guide into a Parlour richly furnished; but where I found nothing fit to be eaten. No Victuals appearing, when I was ready to die of Hunger, made me ask him where the Cloath was laid: But I could not hear what he answered, for at that instant Three or Four young Boys, Children of the House, drew near, and with much Civility stript
me

me to the Shirt. This new Ceremony so astonished me, that I durst not so much as ask my Pretty *Valets de Chamber* the cause of it; and I cannot tell how my Guide, who asked me what I would begin with, could draw from me these two Words, *A Potage*; but hardly had I pronounced them, when I smelt the odour of the most agreable Soop, that ever steamed in the rich Gluttons Nose: I was about to rise from my place, that I might trace that delicious Scent to its source, but my Carrier hindered me: Whither are you going, said he, we shall fetch a walk by and by; but now it is time to Eat, make an end of your *Potage*, and then we'll have something else: And where the Devil is the *Potage*? answered I half angry: Have you laid a wager you'll jeer me all this Day? I thought, replied he, that at the Town we came from, you had seen your Master or some Boelse at meal, and that's the reason I told you not, how People feed in this Country. Seeing then you are still ignorant, you must know, that here they live on Steams. The art of Cookery is to shut up in great Vessels, made on purpose, the Exhalations that proceed from the meat, whilst it is a dressing; and when they have provided enough of

several

several sorts, and several tastes, according to the Appetite of those they treat; they open one Vessel where that Steam is kept, and after that another; and so on till all the Company be satisfied.

Unless you have already lived after this manner, you would never think, that the Nose without Teeth and Gullet, can perform the office of the Mouth, in feeding a Man; but I'll make you experience it your self. He had no sooner said so, but I found so many agreable and nourishing Vapours enter the Parlour, one after another, that in less than half a quarter of an Hour I was fully satisfied. When we were got up; This is not a matter, said he, much to be admired at, seeing you cannot have lived so long, and not have observed, that all sorts of Cooks, who eat less than People of another Calling, are nevertheless much Fatter. Whence proceeds that Plumpness, d'ye think, unless it be from the Steams that continually environ them, which penetrate into their Bodies, and fatten them? Hence it is, that the People of this World enjoy a more steady and vigorous Health, by reason that their Food hardly engenders any Excrements, which are in a manner the original of all Diseases. You were, perhaps, surprised, that before supper

per you were stript, since it is a Custom not practised in your Country; but it is the fashion of this, and for this end used, that the Animal may be the more transpirable to the Fumes. Sir, answered I, there is a great deal of probability in what you say, and I have found somewhat of it my self by experience; but I must frankly tell you, That not being able to Unbrute my self so soon, I should be glad to feel something, that my Teeth might fix upon: He promised I should, but not before next Day; because, said he, to Eat so soon after your meal, would breed Crudities. After we had discoursed a little longer, we went up to a Chamber to take our rest; a Man met us on the top of the Stairs, who having attentively Eyed us, led me into a Closet, where the floor was strowed with Orange-Flowers Three Foot thick, and my Spirit into another, filled with Gilly-Flowers and Jessamine: Perceiving me amazed at that Magnificence, he told me, they were the Beds of the Country. In fine, we laid our selves down to rest, in our several Cells, and so soon as I had stretched my self out upon my Flowers, by the light of Thirty large Glow-worms shut up in a Chrystal, (being the only Candles *Charon* uses,) I perceived

ceived the Three or Four Boys, who had stript me before Supper, One tickling my Feet, another my Thighs, the Third my Flanks, and the Fourth my Arms, and all so delicately and daintily, that in less than in a Minute I was fast asleep.

Next Morning by Sun-rising, my Spirit came into my Room, and said to me, Now I'll be as good as my Word, you shall breakfast this Morning, more solidly than you Supped last Night. With that I got up, and he led me by the Hand to a place, at the back of the Garden, where one of the Children of the House stayed for us, with a Piece in his Hand, much like to one of our Fire-Locks. He asked my Guide, if I would have a dozen of Larks, because *Baboons* (one of which he took me to be,) loved to feed on them? I had hardly answered, Yes, when the Fowler discharged a Shot, and Twenty or Thirty Larks fell at our Feet ready Roasted. This, thought I presently with my self, verifies the Proverb in our World, of a Country where Larks fall ready Roasted; without doubt, it has been made by some Body that came from hence. Fall too, fall too, said my Spirit, don't spare; for they have a knack of mingling a certain Composition with their
Powder

Powder and Shot, which Kills, Plucks, Roasts, and Seasons the Fowl all at once. I took up some of them, and eat them upon his word; and to say the Truth, In all my Life time, I never eat any thing so delicious. Having thus Breakfasted, we prepared to be gone, and with a Thousand odd Faces, which they use when they would shew their Love; our Landlord received a Paper from my Spirit. I asked him, if it was a Note for the Reckoning? He replied, No, that all was paid, and that it was a Copy of Verses. How! Verses, said I, are your Inn-Keepers here curious of Rhime then? It's, said he, the Money of the Country, and the charge we have been at here, hath been computed to amount to Three *Couplets*, or Six Verses, which I have given him. I did not fear we should out-run the Constable; for though we should Pamper our selves for a whole Week, we could not spend a *Sonnet*, and I have Four about me, besides Two *Epigrams*, Two *Odes*, and an *Eclogue*. Would to God, said I, it were so in our World; for I know a good many honest Poets there, who are ready to Starve, and who might live plentifully, if that Money would pass in Payment. I farther asked him, If these Verses would always serve, if one Transcribed them? He made answer, No, and so went on. When an Author has Composed any; he carries them to the Mint, where the sworn Poets of the Kingdom sit in Court. There these versifying Officers essay the pieces; and if they be judged Sterling, they are rated not according to their Coyn; that's to say, That a *Sonnet* is not always as good as a *Sonnet*; but according to the intrinsick value of the piece; so that if any one Starve, he must be a Blockhead: For Men of Wit make always good Chear. With Extasie, I was admiring the judicious Policy of that Country, when he proceeded in this manner: There are others who keep Publick-house, after a far different manner: When one is about to be gone, they demand proportionably to the Charges, an Acquittance for the other World; and when that is given them, they write down in a great Register, which they call *Doomsday's Book*, much after this manner. *Item*, The value of so many Verses, delivered such a Day, to such a Person, which he is to pay upon the receipt of this Acquittance, out of his readiest Cash: And when they find themselves in danger of Death, they cause these Registers to be Chopt in pieces, and swallow them down; because they believe,

that

that if they were not thus digested, they would be good for nothing.

This Conversation was no hinderance to our Journey; for my Four-legged Porter jogged on under me, and I rid stradling on his Back. I shall not be particular in relating to you, all the Adventures, that happened to us on our way, till we arrived at length at the Town, where the King holds his Residence. I was no sooner come, but they carryed me to the Palace, where the Grandees received me with more Moderation, than the People had done, as I passed the Streets: But both great and small concluded, That, without doubt, I was the Female of the Queen's little Animal. My Guide was my Interpreter; and yet he himself understood not the Riddle, and knew not what to make of that little Animal of the Queen's; but we were soon satisfied as to that; for the King having some time considered me, ordered it to be brought, and about half an hour after, I saw a company of Apes, wearing Ruffs and Breeches, come in, and amongst them a little Man, almost of my own Built, for he went on Two Legs; so soon as he perceived me, he Accosted me with a *Criado de vuestra merced.* I answered his Greeting, much in the same Terms. But alas! no sooner had they seen us talk together, but they believed their Conjecture to be true; and so, indeed, it seemed; for he of all the By-standers, that past the most favourable Judgment upon us, protested, that our Conversation, was a Chattering we kept for Joy at our meeting again. That little Man told me, that he was an *European*, a Native of old *Castille*: That he had found a means by the help of Birds, to mount up to the World of the Moon, where then we were: That falling into the Queen's Hands, she had taken him for a Monkey, because Fate would have it so: That in that Country they cloath Apes in a *Spanish* Dress; and that upon his arrival, being found in that habit, she had made no doubt, but he was of the same kind. It could not otherwise be, replied I, but having tried all Fashions of Apparel upon them, none were found so Ridiculous, and by consequence more becoming a kind of Animals, which are only entertained for Pleasure and Diversion. That shews you little understand the Dignity of our Nation, answered he, for whom the Universe breeds Men, only to be our Slaves, and Nature produces nothing but objects of Mirth and Laughter. He then intreated me to tell him, how I durst be so bold, as to Scale the Moon

with

with the Machine I told him of? I answered, That it was because he had carried away the Birds, which I intended to have made use of. He smiled at this Raillery; and about a quarter of an hour after, the King commanded the Keeper of the Monkeys to carry us back, with express Orders to make the *Spaniard* and me lie together, that we might procreate a breed of Apes in his Kingdom. The King's Pleasure was punctually obeyed; at which I was very glad, for the satisfaction I had, of having a Mate to converse with, during the solitude of my Brutification. One Day my Male (for I was taken for the Female) told me, That the true reason, which had obliged him to travel all over the Earth, and at length to abandon it for the Moon, was, that he could not find so much as one Country, where even Imagination was at liberty. Look ye, said he, how the Wittiest thing you can say, unless you wear a Cornered Cap, if it thwart the Principles of the Doctors of the Robe, you are an Ideot, a Fool, and something worse, perhaps. I was about to have been put into the Inquisition at home, for maintaining to the Pedants Teeth, That there was a *Vacuum*, and that I knew no one matter in the World, more Ponderous than another. I asked him, what probable

Arguments he had, to confirm so new an Opinion? To evince that, answered he, you must suppose that there is but one Element; for though we see Water, Earth, Air and Fire distinct, yet are they never found to be so perfectly pure, but that there still remains some Mixture. For example, When you behold Fire, it is not Fire but Air much extended; the Air is but Water much dilated; Water is but liquified Earth, and the Earth it self, but condensed Water; and thus if you weigh Matter seriously, you'll find it is but one, which like an excellent Comedian here below acts all Parts, in all sorts of Dresses: Otherwise we must admit as many Elements, as there are kinds of Bodies: And if you ask me why Fire burns, and Water cools, since it is but one and the same matter, I answer, That that matter acts by Sympathy, according to the Disposition it is in, at the time when it acts. Fire which is nothing but Earth also, more dilated than is fit for the constitution of Air, strives to change into it self, by Sympathy, what ever it meets with: Thus the heat of Coals, being the most subtile Fire, and most proper to penetrate a Body, at first slides through the pores of our Skin; and because it is a new matter that fills us, it makes us exhale in Sweat;

that

that Sweat dilated by the Fire is converted to a Steam, and becomes Air; that Air being farther rarified by the heat of the *Antiperistasis*, or of the Neighbouring Stars, is called Fire, and the Earth abandoned by the Cold and Humidity, which were Ligaments to the whole, falls to the ground: Water, on the other hand, though it no ways differ from the matter of Fire, but in that it is closer, burns us not; because that being dense by Sympathy, it closes up the Bodies it meets with, and the Cold we feel is no more, but the effect of our Flesh contracting it self, because of the Vicinity of Earth or Water, which constrains it to a Resemblance. Hence it is, that those who are troubled with a Dropsie, convert all their nourishment into Water; and the Cholerick convert all the Blood, that is formed in their Liver, into Choler. It being then supposed, that there is but one Element; it is most certain, that all Bodies, according to their several qualities, incline equally towards the Center of the Earth.

But you'll ask me, Why then does Iron, Metal, Earth and Wood, descend more swiftly to the Center than a Sponge, if it be not that it is full of Air, which naturally tends upwards? That is not at all the Reason, and thus I make

make it out: Though a Rock fall with greater Rapidity than a Feather, both of them have the same inclination for the Journey; but a Cannon Bullet, for instance, where the Earth pierced through, would precipitate with greater haste to the Center thereof, than a Bladder full of Wind; and the reason is, because that mass of Metal, is a great deal of Earth contracted into a little space, and that Wind a very little Earth in a large space: For all the parts of Matter, being so closely joined together in the Iron, encrease their force by their Union; because being thus compacted, they are many that Fight against a few, seeing a parcel of Air equal to the Bullet in Bigness, is not equal in Quantity.

Not to insist on a long Deduction of Arguments to prove this, tell me in good earnest, How a Pike, a Sword or a Dagger wound us? If it be not, because the Steel, being a matter, wherein the parts are more continuous, and more closely knit together, than your Flesh is, whose Pores and Softness shew, that it contains but very little Matter, within a great extent of Place; and that the point of the Steel that pricks us, being almost an innumerable number of Particles of matter, against a very little Flesh, it forces it to yeild to the stronger,

in

World of the Moon.

in the same manner as a Squadron in close order, will easily break through a more open Battallion; for why does a Bit of red hot Iron, burn more than a Log of Wood all on Fire? Unless it be, that in the Iron, there is more Fire in a small space, seeing it adheres, to all the parts of the Metal, than in the Wood which being very Spongy, by consequence contains a great deal of *Vacuity*; and that *Vacuity*, being but a Privation of Being, cannot receive the form of Fire. But, you'll object, you suppose a *Vacuum*, as if you had proved it, and that's begging of the question: Well then I'll prove it, and though that difficulty be the Sister of the *Gordian knot*, yet my Arms are strong enough to become its *Alexander*.

Let that vulgar Beast, then, who does not think it self a Man, had it not been told so, answer me if it can: Suppose now there be but one Matter, as I think I have sufficiently proved; whence comes it, that according to its Appetite, it enlarges or contracts its self; whence is it, that a piece of Earth, by being Condensed becomes a Stone? Is it that the parts of that Stone are placed one with another, in such a manner, that wherever that grain of Sand is settled, even there, or in the same point, another grain of Sand is Lodged? That

cannot

cannot be, no not according to their own Principles, seeing there is no Penetration of Bodies: But that matter must have crowded together, and if you will, abridged it self, so that it hath filled some place which was empty before. To say, that it is incomprehensible, that there should be a Nothing in the World, that we are in part made up of Nothing: Why not, pray? Is not the whold World wrapt up in Nothing? Since you yield me this point, then confess ingeniously, that it's as rational, that the World should have a Nothing within it, as Nothing about it.

I well perceive you'll put the question to me, Why Water compressed in a Vessel by the Frost should break it, if it be not to hinder a Vacuity? But I answer, That that only happens, because the Air overhead, which as well as Earth and Water, tends to the Center, meeting with an empty Tun by the way, takes up his Lodging there: If it find the pores of that Vessel, that's to say, the ways that lead to that void place, too narrow, too long, and too crooked, with impatience it breaks through and arrives at its Tun.

But not to trifle away time, in answering all their objections, I dare be bold to say, That if there were no *Vacuity*, there could be no Motion; or else a Penetration of Bodies

dies muſt be admitted; for it would be a little too ridiculous to think, that when a Gnat puſhes back a parcel of Air with its Wings, that parcel drives another before it, that other another ſtill; and that ſo the ſtirring of the little Toe of a Flea, ſhould raiſe a bunch upon the Back of the Univerſe. When they are at a ſtand, they have recourſe to Rarefaction: But in good earneſt, How can it be when a Body is rarified, that one Particle of the Maſs does recede from another Particle, without leaving an empty Space betwixt them; muſt not the two Bodies, which are juſt ſeparated, have been at the ſame time in the ſame place of this; and that ſo they muſt have all three penetrated each other? I expect you'll ask me, why through a Reed, a Syringe or a Pump, Water is forced to aſcend contrary to its inclination? To which I anſwer, That that's by violence, and that it is not the fear of a *Vacuity*, that turns it out of the right way; but that being linked to the Air by an imperceptible Chain, it riſes when the Air, to which it is joined, is raiſed.

That's no ſuch knotty Difficulty, when one knows the perfect Circle, and the delicate Concatenation of the Elements: For if you attentively conſider the Slime, which joines the Earth and Water together in Marriage,

riage, you'll find that it is neither Earth nor Water; but the Mediator betwixt these Two Enemies. In the same manner, the Water and Air reciprocally send a Mist, that dives into the Humours of both, to negotiate a Peace betwixt them; and the Air is reconciled to the Fire, by means of an interposing Exhalation which Unites them.

I believe he would have proceeded in his Discourse, had they not brought us our Victuals; and seeing we were a hungry, I stopt my Ears to his discourse, and opened my Stomack to the Food they gave us.

I remember another time, when we were upon our Philosophy; for neither of us took pleasure to Discourse of mean things: I am vexed, said he, to see a Wit of your stamp, infected with the Errors of the Vulgar. You must know then, in spight of the Pedantry of *Aristotle*, with which your Schools in *France* still ring, That every thing is in every thing; that's to say, for instance, That in the Water there is Fire, in the Fire Water, in the Air Earth, and in the Earth Air: Though that Opinion makes Scholars open their Eyes as big as Sawcers, yet it is easier to prove it, than perswade it. For I ask them, in the first place, if Water does not breed Fish: If they deny it, let them dig a Pit, fill it with meer Element,

ment, and to prevent all blind Objections, let them if they please, strain it through a Strainer, and I'll oblige my self, in case they find no Fish therein, within a certain time, to drink up all the Water they have poured into it: But if they find Fish, as I make no doubt on't; it is a convincing Argument, that there is both Salt and Fire there. Consequentially now, to find Water in Fire; I take it to be no difficult Task. For let them chuse Fire, even that which is most abstracted from Matter, as Comets are, there is a great deal in them still; seeing if that Unctuous Humour, whereof they are engendred, being reduced to a Sulphur, by the heat of the Antiperistasis which kindles them, did not find a curb of its Violence in the humid Cold, that qualifies and resists it, it would spend it self in a trice like Lightning. Now that there is Air in the Earth, they will not deny it; or otherwise they have never heard of the terrible Earth-quakes, that have so often shaken the Mountains of *Sicily:* Besides, the Earth is full of Pores, even to the least grains of Sand that compass it. Nevertheless, no Man hath as yet said, that these Hollows were filled with *Vacuity*: It will not be taken amiss then, I hope, if the Air takes up its quarters there. It remains to be proved, that there is Earth in the

the Air; but I think it scarcely worth my pains, seeing you are convinced of it, as often as you see, such numberless Legions of Atomes fall upon your heads, as even stiffle Arithmetick.

But let us pass from simple to compound Bodies, they'll furnish me with much more frequent Subjects; and to demonstrate that all things are in all things, not that they change into one another, as your *Peripateticks* Juggle, (for I will maintain to their Teeth, that the Principles mingle, separate, and mingle again in such a manner, that that hath been made Water by the Wise Creator of the World, will always be Water:) I shall suppose no Maxime, as they do, but what I prove.

And therefore take a Billet, or any other combustible stuff, and set Fire to it, they'll say when it is in a Flame, That what was Wood is now become Fire; but I maintain the contrary, and that there is no more Fire in it, when it is all in Flame, than before it was kindled; but that which before was hid in the Billet, and by the Humidity and Cold hindered from acting; being now assisted by the Stronger, hath rallied its forces against the Phlegm that choaked it, and commanding the Field of Battle, that was possessed by its Enemy, triumphs over his Jaylor, and appears without Fetters.

ters. Don't you see how the Water flees out at the two ends of the Billet, hot and smoaking from the Fight it was engaged in. That flame which you see rise on high, is the purer Fire, unpestered from the Matter, and by consequence the readiest to return home to it self: Nevertheless it Unites it self, by tapering into a Piramide, till it rise to a certain height, that it may pierce through the thick Humidity of the Air, which resists it; but as in mounting it disengages it self by little and little, from the violent company of its Landlords; so it diffuses it self, because then it meets with nothing that thwarts its passage, which negligence, though, is many times the cause of a second Captivity: For marching stragglingly, it wanders sometimes into a Cloud, and if it meet there with a Party of its own, sufficient to make head against a Vapour; they Engage, Grumble, Thunder and Roar, and the Death of Innocents is many times the effect of the animated Rage, of those inanimated Things. If, when it finds it self pestered, among those Crudities of the middle Region, it is not strong enough to make a defence; it yields to its Enemy upon discretion, which by its weight, constrains it to fall again to the Earth: And this Wretch, inclosed in a drop of Rain, may, perhaps, fall at the Foot of an

F Oak,